BLIND ESTIMATION USING HIGHER-ORDER STATISTICS

BLIND ESTIMATION USING HIGHER-ORDER STATISTICS

Edited by

Asoke Kumar Nandi

David Jardine Chair of Electrical Engineering,
Department of Electrical Engineering and Electronics,
University of Liverpool

KLUWER ACADEMIC PUBLISHERS

BOSTON / DORDRECHT / LONDON

A C.I.P. Catalogue record for this book is available from the Library of Congress.

ISBN 978-1-4419-5078-9

Published by Kluwer Academic Publishers,
P.O. Box 17, 3300 AA Dordrecht, The Netherlands.

Sold and distributed in North, Central and South America
by Kluwer Academic Publishers,
101 Philip Drive, Norwell, MA 02061, U.S.A.

In all other countries, sold and distributed
by Kluwer Academic Publishers,
P.O. Box 322, 3300 AH Dordrecht, The Netherlands.

Printed on acid-free paper

To my family
Marion, Robin, David, and Anita Nandi,
for their sacrifice, support and love.

Contents

1 Higher–Order Statistics   **1**
A McCormick and A K Nandi

1.1 Introduction 2
1.2 Stochastic Processes 2
1.3 Moments and Cumulants 4
1.4 Pictorial Motivation for HOS 12
1.5 Minimum and Nonminimum Phase Systems 14
1.6 Cyclostationary Signals and Statistics 19
1.7 Estimation of Cyclic-statistics 21
1.8 Summary 23
References 25

2 Blind Signal Equalisation   **27**
S N Anfinsen, F Herrmann and A K Nandi

2.1 Introduction 29
2.2 Gradient Descent Algorithms 40
2.3 Blind Equalisation Algorithms 42
2.4 Algorithms Based on Explicit HOS 55
2.5 Equalisation with Multiple Channels 67
2.6 Algorithms Based on Cyclostationary Statistics 84
2.7 General Convergence Considerations 93
2.8 Discussion 95
References 97

3 Blind System Identification   **103**
J K Richardson and A K Nandi

3.1 Introduction 105
3.2 MA Processes 108
3.3 ARMA Processes 136
References 162

4 Blind Source Separation   **167**
   V Zarzoso and A K Nandi

   4.1 Introduction 169
   4.2 Problem statement 171
   4.3 Separation quality: performance indices 185
   4.4 A real-life problem: the fetal ECG extraction 188
   4.5 Methods based on second-order statistics 189
   4.6 Methods based on higher-order statistics 203
   4.7 Comparison 231
   4.8 Comments on the literature 236
   References 247

5 Robust Cumulant Estimation   **253**
   D Mämpel and A K Nandi

   5.1 Introduction 254
   5.2 AGTM, LMS and LTS 255
   5.3 The $q_0 - q_2$ plane 257
   5.4 Continuous probability density functions 257
   5.5 Algorithm 259
   5.6 Simulations and results 263
   5.7 Concluding Remarks 273
   References 275

Epilogue   **279**

Index   **280**

Preface

Higher-order statistics (HOS) is itself an old subject of enquiry. But in signal processing research community, a number of significant developments in higher-order statistics begun in mid-1980's. Since then every signal processing conference proceedings contained papers on HOS. The IEEE has been organising biennial workshops in HOS since 1989. There have been many Special Issues on HOS in various journals - including 'Applications of Higher-Order Statistics' (ed. J M Mendel and A K Nandi), IEE Proceedings, Part F, vol. 140, no. 6, pp. 341-420 and 'Higher-Order Statistics in Signal Processing' (ed. A K Nandi), Journal of the Franklin Institute, vol. 333B, no. 3, pp. 311 452.

These last fifteen years have witnessed a large number of theoretical developments as well as real applications. There are available very few books in the subject of HOS and there are no books devoted to blind estimation. Blind estimation is a very interesting, challenging and worthwhile topic for investigation as well as application. The need for a book covering both HOS and blind estimation has been felt for a while. Thus the goal in producing this book has been to focus in the blind estimation area and to record some of these developments in this area. This book is divided into five main chapters. The first chapter offers an introduction to HOS; more on this may be gathered from the existing literature. The second chapter records blind signal equalisation which has many applications including (mobile) communications. A number of new and recent developments are detailed therein. The third chapter is devoted to blind system identification. Some of the published algorithms are presented in this chapter. The fourth chapter is concerned with blind source separation which is a generic problem in signal processing. It has many applications including radar, sonar, and communications. The fifth chapter is devoted to robust cumulant estimation. This chapter is primarily based on ideas and experimental work with little solid theoretical foundation but the problem is an important one and results are encouraging. It deserves more attention and hopefully that will now be forthcoming.

All such developments are still continuing and therefore a book, such as this one, cannot be definitive or complete. It is hoped however that it will fill an important gap; students embarking on graduate studies should be able to learn enough basics before tackling journal papers, researchers in related fields should be able to get a broad perspective on what has been achieved, and current researchers in the field should be able to use it as some kind of reference. The subject area has been introduced, some major developments have been recorded, and enough success as well as challenges are noted here for more people to look into higher-order statistics, along with any other information, for either generating solutions of problems or solutions of their own problems.

I wish to acknowledge the efforts of all the contributors, who have worked very hard to make this book possible. A work of this magnitude will unfortunately contain errors and omissions. I would like to take this opportunity to apologise unreservedly for all such indiscretions. I would welcome comments or corrections; please send them to me by email (a.k.nandi@ieee.org) or any other means.

Asoke K Nandi
Glasgow, UK
October 1998

1 HIGHER-ORDER AND CYCLOSTATIONARY STATISTICS

A McCormick and A K Nandi

Contents

1.1	Introduction	2
1.2	Stochastic Processes	2
1.3	Moments and Cumulants	4
	1.3.1 Definitions	5
	1.3.2 Salient Cumulants Properties	6
	1.3.3 Moment and Cumulant Estimation	7
	1.3.4 Spectral Estimation	10
	1.3.5 Estimation of Power Spectra	10
	1.3.6 Estimation of Bispectra	11
1.4	Pictorial Motivation for HOS	12
1.5	Minimum and Nonminimum Phase Systems	14
	1.5.1 Minimum Phase Systems	14
	1.5.2 Nonminimum Phase systems	14
	1.5.3 Phase Blindness of Second-Order Statistics	17
1.6	Cyclostationary Signals and Statistics	19
1.7	Estimation of Cyclic-statistics	21
1.8	Summary	23
	References	25

1.1   Introduction

Until the mid-1980's, signal processing - signal analysis, system identification, signal estimation problems, etc. - was primarily based on second-order
statistical information. Autocorrelations and cross-correlations are examples
of second-order statistics (SOS). The power spectrum which is widely used
and contains useful information is again based on the second-order statistics in that the power spectrum is the one-dimensional Fourier transform of
the autocorrelation function. As Gaussian processes exist and a Gaussian
probability density function (pdf) is completely characterised by its first two
moments, the analysis of linear systems and signals has so far been quite
effective in many circumstances. It has nevertheless been limited by the
assumptions of Gaussianity, minimum phase systems, linear systems, etc.

Another common major assumption in signal processing is that of signal
ergodicity and stationarity. These assumptions allow the statistics and other
signal parameters of the signal to be estimated using time averaging. However
in some cases the signal parameters being estimated are changing with time
and therefore estimates based on these assumptions will not provide accurate
parameter estimates. Non-stationary signals cannot be characterised using
these traditional approaches. One specific case of non-stationarity is that of
cyclo-stationarity. These signals have statistics which vary periodically.

1.2   Stochastic Processes

A stochastic process is a process which takes the value of a series of random
variables over time, e.g. $X(t)$. Random variables do not have a predictable
value, but the probability of a random variable, X, taking a particular value,
x, is determined by its probability density function, $p(x)$, and this can be estimated from an ensemble of samples of the random variable, $\{x_1, x_2, \ldots, x_n\}$.
In many cases the probability density function, and hence the behaviour of
the random variable, can be characterised by a few statistical parameters
such as moments or cumulants, e.g. the mean, $\mu = \frac{1}{N} \sum_{k=1}^{N} x_k$. For a Gaussian random variable, the first two cumulants, the mean (μ) and variance
(σ^2) are sufficient to characterise the pdf:

$$p(x) = \frac{1}{\sqrt{2\pi}} \exp\left(\frac{-(x-\mu)^2}{2\sigma^2}\right) \tag{1.1}$$

A stochastic process is a series of these random values occurring at successive
points in time, $\{X(t_1), X(t_2), \ldots, X(t_n)\}$. If the pdf, $p(x, t)$ of each random
variable in the time series is identical then the process is said to be stationary,

i.e. $p(x, t_1) = p(x, t_2) \forall t_1, t_2$. The pdf can depend on previous process values, or the random variables can be independent in which case the process is termed white and its power spectrum is flat.

Statistical signal processing treats the sampled signal as a stochastic process. The underlying physical process being measured may be deterministic or stochastic. Measurement errors in sampling the signal may also produce stochastic components in the signal. The stochastic signal can be characterised by statistical parameters such as moments or by spectral parameters. Spectral parameters are popular as they relate to the Fourier decomposition of deterministic periodic signals, and are especially useful if the underlying process is deterministic and periodic.

An important concept in statistical signal processing is that of ergodicity. This means that statistical averages can be equated to time averages, e.g. with the signal mean:

$$\mu = \frac{1}{N} \sum_{k=1}^{N} x_k = \frac{1}{T} \int_0^T X(t) dt \qquad (1.2)$$

When determining the moments of signals, such as the mean value (the first order moment), every sample in the signal must therefore have the same distribution, and hence the signal must be stationary. The power spectrum is often estimated using the averaged periodogram approach, where the power spectrum is estimated as the average magnitude of the Fourier transform of sections of the signal taken over separate sections.

$$S_{XX}(\omega) = \frac{1}{N} \sum_{k=1}^{N} |F_k(\omega)| \qquad (1.3)$$

where N sections of the signal have been used to estimate the Fourier transform $F_k(\omega) = \int_{kT-T/2}^{kT+T/2} X(t) e^{-\jmath \omega t} dt$.

Ergodicity applies here when the magnitudes of the successive Fourier Transforms have the same pdf, $p(|F_1(\omega)|) = p(|F_2(\omega)|) = \cdots = p(|F_k(\omega)|)$. The ergodicity therefore applies to all stationary signals. It also applies to all periodic signals. If the phase of a periodic signal is known then, successive samples are predictable and have different pdfs and are therefore non-stationary. However periodic signals with random phase are stationary, as the pdf of all the samples in the signal is the same, and this random phase is often introduced as an effect of the random time at which the signal sampling starts. Many signals are non-stationary, with their moments, pdfs and even spectral characteristics changing over time. One significant class of non-stationary signals are cyclostationary signals. These signals have the property that samples separated by a period have the same pdf. This opens up the opportunity to exploit samples separated by the cycle-period to create an ensemble of points and thus better estimate signal characteristics.

1.3   Moments and Cumulants

When signals are non-Gaussian the first two moments do not define their pdf and consequently higher-order statistics (HOS), namely of order greater than two, can reveal other information about them than SOS alone can. Ideally the entire pdf is needed to characterise a non-Gaussian signal. In practice this is not available but the pdf may be characterised by its moments. It should however be noted that some distributions do not possess finite moments of all orders. As an example, Cauchy distribution, defined as

$$p(x) = \frac{1}{\pi \beta} \ \frac{1}{1 + (\frac{x-\alpha}{\beta})^2} \ , \qquad -\infty < x < \infty \tag{1.4}$$

has all its moments, including the mean, undefined. Also some distributions give rise to finite moments but these moments do not uniquely define the distributions. For example, the log-normal distribution is not determined by its moments [9]. As an example of the fact that different distributions can have the same set of moments, consider

$$dF(x) = \gamma \exp(-\alpha x^\lambda) \left\{ 1 + \epsilon \sin(\beta x^\lambda) \right\} \ dx, \tag{1.5}$$

for $0 \leq x \leq \infty$, $\alpha > 0$, $0 < \lambda < 1/2$, and $|\epsilon| < 1$. The interesting thing about this set of distributions (obtained for different values of ϵ) is that they all have the same set of moments for all allowed values of ϵ in the range $|\epsilon| < 1$ [20] because

$$\int_0^\infty x^n \exp(-\alpha x^\lambda) \sin(\beta x^\lambda) \ dx = 0 \ . \tag{1.6}$$

Thus it is clear that the moments, even when they exist for all orders, do not necessarily determine the pdf completely. Only under certain conditions will a set of moments determine a pdf uniquely. It is rather fortunate that these conditions are satisfied by most of the distributions arising commonly. For practical purposes, the knowledge of moments may be considered equivalent to the knowledge of the pdf. Thus distributions that have a finite number of the lower moments in common will, in a sense, be close approximations to each other. In practice, approximations of this kind often turn out to be remarkably good, even when only the first three or four moments are equated [18].

1.3.1 Definitions

Let the cumulative distribution function (cdf) of x be denoted by $F(x)$. The central moment (about the mean) of order ν of x is defined by

$$\mu_\nu = \int_{-\infty}^{\infty} (x - m)^\nu \, dF \qquad (1.7)$$

for $\nu = 1, 2, 3, 4, \dots$ where m, the mean of x, is given by $\int_{-\infty}^{\infty} x \, dF$, $\mu_0 = 1$ and $\mu_1 = 0$. As noted earlier, not all distributions have finite moments of all orders; for example, the Cauchy distribution belongs to this class. In the following it is assumed that distributions are zero-mean. One can also introduce the characteristic function, for real values of t,

$$\phi(t) = \int_{-\infty}^{\infty} \exp(\jmath t x) \, dF = \sum_{\nu=0}^{\infty} \mu_\nu (\jmath t)^\nu / \nu!, \qquad (1.8)$$

where $\jmath = \sqrt{-1}$ and μ_ν is the moment of order ν about the origin. Hence coefficients of $(\jmath t)^\nu / \nu!$ in the power series expansion of the $\phi(t)$ represent moments. Moments are thus one set of descriptive constants of a distribution. In general, moments may not completely determine the distribution even when moments of all orders exist. For example, the log-normal distribution is not uniquely determined by its moments.

Cumulants make up another set of descriptive constants. If one were to express $\phi(t)$ as follows,

$$\phi(t) = \int_{-\infty}^{\infty} \exp\left(\jmath t x\right) dF) = \exp\left(\sum_{\nu=1}^{\infty} C_\nu (\jmath t)^\nu / \nu! \right), \qquad (1.9)$$

then the C_ν's are the cumulants of x and these are the coefficients of $(\jmath t)^\nu / \nu!$ in the power series expansion of the natural logarithm of $\phi(t)$, $\ln \phi(t)$. The cumulants, except for the C_1, are invariant under the shift of the origin, a property that is not shared by the moments.

Cumulants and moments are different though clearly related (as seen through the characteristic function). Cumulants are not directly estimable by summatory or integrative processes, and to find them it is necessary either to derive them from the characteristic function or to find the moments first. For zero-mean distributions, the first three central moments and the corresponding cumulants are identical but they begin to differ from order four — i.e. $C_1 = \mu_1 = 0$, $C_2 = \mu_2$, $C_3 = \mu_3$, and $C_4 = \mu_4 - 3\mu_2^4$. For zero-mean Gaussian distributions, $C_1 = 0$ (zero-mean), $C_2 = \sigma^2$ (variance), and $C_\nu = 0$

for $\nu > 2$. On the other hand for Poisson distributions, $C_\nu = \lambda$ (mean) for all values of ν.

For a zero-mean, real, stationary time-series $\{x(k)\}$ the second-order moment sequence (autocorrelations) is defined as

$$M_2(k) = M_{xx}(k) = \mathcal{E}\{x(i)x(i+k)\} \tag{1.10}$$

where $E[\cdot]$ is the expectation operator and i is the time index. In this case the second-order cumulants, $C_2(k)$, are the same as $M_2(k)$, i.e. $C_2(k) = C_{xx}(k) = M_2(k) \ \forall \ k$. The third-order moment sequence is defined by

$$M_3(k,m) = M_{xxx}(k,m) = E[x(i)x(i+k)x(i+m)] \tag{1.11}$$

and again $C_3(k,m) = C_{xxx}(k,m) = M_3(k,m) \ \forall \ k,m$ where $C_3(.,.)$ is the third-order cumulant sequence. The fourth-order moment sequence is defined as

$$M_4(k,m,n) = M_{xxxx}(k,m,n) = E[x(i)x(i+k)x(i+m)x(i+n)] \tag{1.12}$$

and the fourth-order cumulants are

$$\begin{aligned}
C_4(k,m,n) &= C_{xxxx}(k,m,n) \\
&= M_4(k,m,n) - C_2(k)C_2(m-n) - C_2(m)C_2(k-n) \\
&\quad - C_2(n)C_2(m-k)
\end{aligned}$$

As can be seen the fourth-order moments are different from the fourth-order cumulants.

1.3.2 Salient Cumulants Properties

Although the moments of a system provide all the information required for analysis of a random process it is usually more preferable to work with related quantities called cumulants which more clearly exhibit the additional information included using higher-order statistics. Their use is analogous to using the covariance instead of the correlation function in second moment analysis to remove the effect of the mean. Higher-order cumulants measure the departure of a random process from a Gaussian random process with an identical mean and covariance function. Thus Gaussian random processes have higher-order cumulants which are identically zero. In addition, if two or more sets of random variables $\{x[1], x[2], \ldots, x[K]\}$ and $\{v[1], v[2], \ldots, v[K]\}$ are statistically independent then the l-th order cumulant of the random variable $y[k] = x[k] + v[k]$ is equal to the sum of the l-th order cumulants of the two independent sequences

$$C_{ly}(\tau_1, \tau_2, \ldots, \tau_{l-1}) = C_{lx}(\tau_1, \tau_2, \ldots, \tau_{l-1}) + C_{lv}(\tau_1, \tau_2, \ldots, \tau_{l-1}) . \quad (1.13)$$

This is not the case for higher-order moments

$$M_{ly}(\tau_1, \tau_2, \ldots, \tau_{l-1}) \neq M_{lx}(\tau_1, \tau_2, \ldots, \tau_{l-1}) + M_{lv}(\tau_1, \tau_2, \ldots, \tau_{l-1}) . \quad (1.14)$$

Thus, given a non-Gaussian signal that is corrupted by additive Gaussian noise the use of higher-order cumulants theoretically results in elimination of the additive Gaussian noise. This feature of higher-order cumulants means that any estimation of a Gaussian corrupted signal using higher-order cumulants results in automatic noise reduction which can be exploited in the blind (linear) system identification techniques described in chapters 3, 4 and 5 of this thesis. Cumulants have many additional properties which can be exploited to reduce estimation costs. These are described fully in [16]. The symmetry properties of the cumulants of real random processes are given below

$$
\begin{aligned}
C_{3x}(\tau_1, \tau_2) &= C_{3x}(\tau_2, \tau_1) = C_{3x}(-\tau_2, \tau_1 - \tau_2) & (1.15) \\
&= C_{3x}(-\tau_1, \tau_2 - \tau_1) = C_{3x}(\tau_2 - \tau_1, -\tau_1) & (1.16) \\
&= C_{3x}(\tau_1 - \tau_2, \tau_2) . & (1.17)
\end{aligned}
$$

Thus for real random processes, the estimation of cumulants in the region defined by $\tau_1 = 0, \tau_2 \leqslant \tau_1$ is sufficient to define the cumulant sequence, thereby reducing computational requirements.

1.3.3 Moment and Cumulant Estimation

In practice, a finite number of data samples are available - $\{x(i), i = 1, 2, \ldots, N\}$. These are assumed to be samples from a real, zero-mean, stationary process. The sample estimates at second-order are given by

$$\hat{M}_2(k) = \frac{1}{N_3} \sum_{i=N_1}^{N_2} x(i)x(i+k) \quad (1.18)$$

and

$$\hat{C}_2(k) = \hat{M}_2(k) \quad (1.19)$$

where

$$|k| < N\,, \quad N_1 = \begin{cases} 1\,, & \text{if} \quad k \geq 0 \\ -k+1\,, & \text{if} \quad k < 0 \end{cases}\,, \quad N_2 = \begin{cases} N-k\,, & \text{if} \quad k \geq 0 \\ N\,, & \text{if} \quad k < 0 \end{cases}$$

If N_3 is set to the actual number of terms in the summation, namely ($N_2 - N_1 + 1$), unbiased estimates are obtained. Usually N_3 is set to N, the number of data samples, to obtain asymptotically unbiased estimates. Similarly sample estimates of third-order moments and cumulants are given by

$$\hat{M}_3(k,m) = \frac{1}{N_3} \sum_{i=N_1}^{N_2} x(i)x(i+k)x(i+m) \tag{1.20}$$

and

$$\hat{C}_3(k,m) = \hat{M}_3(k,m) \tag{1.21}$$

where N_1 and N_2 take up different values from those in the second-order case. Such estimates are known to be consistent under some weak conditions. For large sample numbers N, the variance of the third-order cumulants can be expected as follows

$$\text{var}\,\{\hat{C}_3(k,m)\} \propto \frac{1}{N}. \tag{1.22}$$

Finally the fourth-order moments are estimated by

$$\hat{M}_4(k,m,n) = \frac{1}{N_3} \sum_{i=N_1}^{N_2} x(i)x(i+k)x(i+m)x(i+n) \tag{1.23}$$

where N_1 and N_2 take up different values from those in the second-order as well as third-order cases and the fourth-order cumulants can be written as

$$\hat{C}_4(k,m,n) = \hat{M}_4(k,m,n) - \hat{M}_2(k)\hat{M}_2(m-n) - \hat{M}_2(m)\hat{M}_2(k-n) \\ - \hat{M}_2(n)\hat{M}_2(m-k) \tag{1.24}$$

As these assume that the processes are zero-mean, in practice the sample mean is removed before calculating moments and cumulants. Mean square convergence and asymptotic normality of the sample cumulant estimates under some mixing conditions are given in [3].

Thus standard estimation method evaluates third-order moments as

$$\hat{M}_3(k,m) = \frac{1}{N_3} \sum_{i=N_1}^{N_2} x(i)x(i+k)x(i+m) = \frac{1}{N_3} \sum_{i=N_1}^{N_2} z_{k,m}(i) \tag{1.25}$$

where $z_{k,m}(i) \equiv x(i)x(i+k)x(i+m)$. This last formulation demonstrates that the standard evaluation employs the mean estimator (of $z_{k,m}(i)$). Sometimes the time-series data are segmented and the set of required cumulants in each of these segments is estimated separately using the mean estimator, and then for the final estimate of a cumulant the mean of the same is calculated over all the segments. Accuracy of methods based on higher-order cumulants depends on, among others, the accuracy of estimates of the cumulants. By their very nature, estimates of third-order cumulants of a given set of data samples tend to be more variable than the autocorrelations (second-order cumulants) of the data. Any error in the values of cumulants estimated from finite segments of a time-series will be reflected as larger variance in other higher-order estimates.

Numerous algorithms employing HOS have been proposed for applications in areas such as array processing, blind system identification, time-delay estimation, blind deconvolution and equalisation, interference cancellation, etc. Generally these use higher-order moments or cumulants of a given set of data samples. One of the difficulties with HOS is the increased computational complexity. One reason is that, for a given number of data samples, HOS computation requires more multiplications than the corresponding SOS calculation.

Another important reason lies in the fact that, for a given number of data samples, variances of the higher-order cumulant estimates are generally larger than that of the second-order cumulant estimates. Consequently, to obtain estimates of comparable variance, one needs to employ a greater number of samples for HOS calculations in comparison to SOS calculations. Using a moderate number of samples, standard estimates of such cumulants are of comparatively high variance and to make these algorithms practical one needs to obtain some lower variance sub-asymptotic estimates.

Recently the problem of robust estimation of second and higher-order cumulants has been addressed [1, 2, 10, 11, 14]. The mean estimator is the one utilised in applications to date. A number of estimators including the mean, median, biweight, and wave were compared using random data. It has been argued that, for not too large number of samples, the mean estimator is not optimal and this has been supported by extensive simulations. Also were developed some generalised trimmed mean estimators for moments and these appear to perform better than the standard estimator in small number of samples in simulations as well as in the estimates of the bispectrum using real data [11, 14]. Another important issue relating to the effects of finite register length (quantisation noise) on the cumulant estimates are being considered (see, for example, [10]).

1.3.4 Spectral Estimation

Let a zero-mean, real, stationary time-series $\{x(k)\}$ represent the observed signal. It is well known that the power spectrum of this signal can be defined as the one-dimensional Fourier transform of the autocorrelations (second-order cumulants) of the signal. Therefore,

$$S_2(\omega_1) = \sum_m C_2(m) \exp(-\jmath\omega_1 m), \qquad (1.26)$$

where

$$C_2(m) = \mathrm{E}[x(k)x(k+m)] , \qquad (1.27)$$

and ω_1 is the frequency.

Similarly, the bispectrum (based on the third-order statistics) of the $x(k)$ can be defined as the two-dimensional Fourier transform of the third-order cumulants, i.e.

$$S_3(\omega_1,\omega_2) = \sum_m \sum_n C_3(m,n) \exp(-\jmath(m\omega_1 + n\omega_2)) , \qquad (1.28)$$

where the $C_3(m,n) = \mathcal{E}\{x(k)x(k+m)x(k+n)\}$ is the third-order cumulant sequence. Correspondingly, the trispectrum (based on the fourth-order statistics) of the $\{x(k)\}$ can be defined as the three-dimensional Fourier transform of the fourth-order cumulants, i.e.

$$S_4(\omega_1,\omega_2,\omega_3) = \sum_m \sum_n \sum_l C_4(m,n,l) \exp(-\jmath(m\omega_1 + n\omega_2 + l\omega_3)) , \qquad (1.29)$$

where the $C_4(m,n,l)$ is the fourth-order cumulant sequence.

However, just as the power spectrum can be estimated from the Fourier transform of the signal (rather than its autocorrelations), one can estimate the bispectrum and trispectrum from the same Fourier transform. The difference will be in the variance of the resulting estimates. It should be noted that consistent estimators of higher-order cumulant spectra via spectral windows and the limiting behaviour of certain functionals of higher-order spectra have been studied.

1.3.5 Estimation of Power Spectra

The definition of the power spectrum involves summing over an infinite data length which is obviously impossible in practice. An estimate of the power spectrum can be obtained by assuming a finite data set, $x[1], x[2], \ldots, x[K]$, of K samples and redefining the sample spectrum as

$$S_2(\omega) = \frac{2\pi}{\omega K} \left| \sum_{k=0}^{K} x[k] e^{-j2\pi\omega k} \right|^2 \tag{1.30}$$

However, this approach yields inconsistent estimates. Various averaging schemes have been proposed to eliminate inconsistent estimates. Welch's periodogram method was used in order to obtain consistent and smooth power spectral estimates. [12] describes Welch's method in full. This method allows data segments to overlap, thereby increasing the number of segments that are averaged and decreasing the variance of the power spectral density estimate. All power spectral estimates were implemented using the MATLAB function PSD.M which employs Welch's periodogram method [12]. Details of the segmentation and windowing employed are given along with the power spectral estimates.

1.3.6 Estimation of Bispectra

There are three types of conventional approach for estimating the bispectrum of a finite length time series. An indirect approach was used where estimates of the cumulants are made first and a 2-dimensional Fourier Transform is applied to estimate the bispectrum. Like the power spectrum however, such estimation would result in inconsistent estimates and the data must be segmented and windowed. The properties of suitable window functions for bispectrum estimation are detailed in [16]. One such window function is an extension of the 1-dimensional Parzen window to a 2-dimensional window function. The bispectral estimates were obtained by computing unbiased estimates of third-order cumulants for each record and averaging the cumulant estimates across all records. The 1-dimensional Parzen window defined by equation (1.31) was extended to a 2-dimensional window using equation (1.32) and applied to the cumulant data. The bispectrum was obtained by taking the 2-dimensional Fast Fourier Transform (FFT) of the windowed cumulant function. Full details of the windowing process are given with the bispectrum estimates.

$$d_p(m) = \begin{cases} 1 - 6\left(\frac{|m|}{L}\right)^2 + 6\left(\frac{|m|}{L}\right)^3 & , \text{if} \quad |m| \leq L/2, \\ 2\left(1 - \frac{|m|}{L}\right)^3 & , \text{if} \quad L/2 < m \leq L, \\ 0 & , \text{if} \quad m > L, \end{cases} \tag{1.31}$$

and

$$W(m, n) = d_p(m) d_p(n) d_p(m - n) \tag{1.32}$$

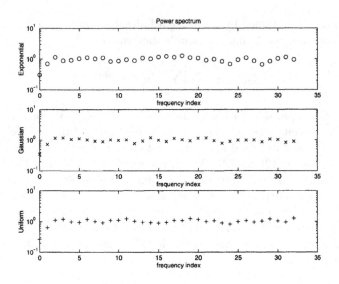

Figure 1.1: Power spectrum of random signals

1.4   Pictorial Motivation for HOS

Three time-series corresponding to independent and identically distributed
(i.i.d.) exponential, Gaussian and uniform random variables (r.v.) are simulated in MATLAB [13] and each of these has 4096 samples of zero mean
and unit variance. Figure 1.1 shows the estimated power spectrum (the top
one corresponds to exponential, the middle one to Gaussian and the bottom
one to uniform) versus the frequency index, while figure 1.2 (o corresponds
to exponential, x corresponds to Gaussian and + corresponds to uniform)
shows the estimated second-order cumulant versus the lag. It is clear that
in both of these two figures, which represent second-order statistical information, exponential, Gaussian and uniform r.v. are not differentiated. Figure
1.3 shows the histograms of these three sets of r.v. from which the differences
are visually obvious. Figure 1.4 shows the estimated third-order cumulants,
$\hat{C}_3(k, k)$, versus the lag, k. In this figure, exponential r.v. are clearly distinguished from Gaussian and uniform r.v. Figure 1.5 shows the estimated
fourth-order cumulants, $\hat{C}_4(k, k, k)$, versus the lag, k. Unlike in the last figure, now it is clear that all three sets of r.v. are differentiated in this figure.
The reasons for the above are obvious from table 1.1, which records theoretical values of cumulants of up to order four for these three types of r.v. of
zero mean and unit variance, and from table 1.2, which presents estimated

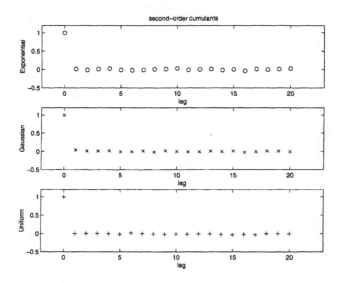

Figure 1.2: Autocorrelations of random signals

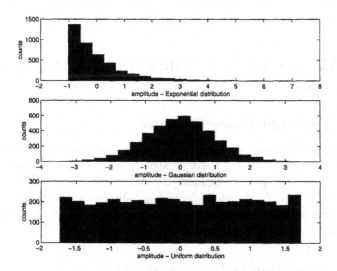

Figure 1.3: Histograms of random signals

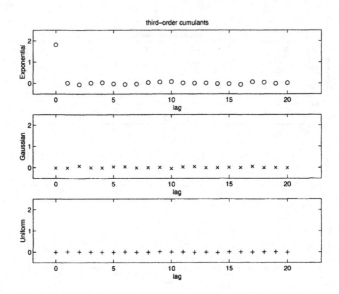

Figure 1.4: Third-order diagonal cumulants of random signals

values of cumulants up to order four at zero lag. In particular all cumulants of i.i.d., Gaussian r.v. beyond order two are theoretically zero.

1.5 Minimum and Nonminimum Phase Systems

1.5.1 Minimum Phase Systems

For a causal discrete system to be described as minimum phase (MP) the zeros of that discrete system must lie strictly inside the unit circle. The transfer function of figure (1.6) must therefore possess a rational transfer function where both $B(z)$ and $A(z)$ are minimum phase polynomials. Such an MP system is a causal stable system and the inverse of such an MP system is causal and stable also.

1.5.2 Nonminimum Phase systems

If all the zeros are outside the unit circle the system is described as maximum phase (MXP); if some of the zeros are inside the unit circle whilst others lie

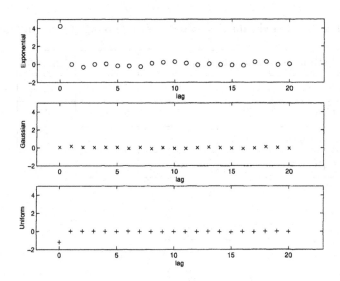

Figure 1.5: Fourth-order diagonal cumulants of random signals

Table 1.1: Theoretical values of cumulants of random signals

	Exponential	Gaussian	Uniform
C_1	0	0	0
$C_2(k)$	$\begin{cases} 1, \text{ for } k = 0 \\ 0, \text{ otherwise} \end{cases}$	$\begin{cases} 1, \text{ for } k = 0 \\ 0, \text{ otherwise} \end{cases}$	$\begin{cases} 1, \text{ for } k = 0 \\ 0, \text{ otherwise} \end{cases}$
$C_3(k, k)$	$\begin{cases} 2, \text{ for } k = 0 \\ 0, \text{ otherwise} \end{cases}$	0	0
$C_4(k, k, k)$	$\begin{cases} 6, \text{ for } k = 0 \\ 0, \text{ otherwise} \end{cases}$	0	$\begin{cases} -1.2, \text{ for } k = 0 \\ 0, \text{ otherwise} \end{cases}$

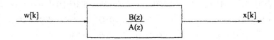

Figure 1.6: System Transfer Function Schematic

Table 1.2: Estimated values of cumulants at zero-lag of random signals

Estimated cumulant	Exponential	Gaussian	Uniform
$\hat{C}_1$	0	0	0
$\hat{C}_2(0)$	1	1	1
$\hat{C}_3(0,0)$	1.8	0	0
$\hat{C}_4(0,0,0)$	4.2	.1	-1.2

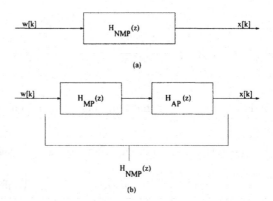

Figure 1.7: (a) Nonminimum Phase System, (b) Representation of NMP System as a MP System Cascaded with an AP System

outside the unit circle the system is a mixed phase system and is described as a nonminimum phase (NMP) system. Since the zeros and poles of the system are interchanged in the system inverse, a nonminimum phase system is either noncausal or unstable. Providing that a system never has any poles or zeros precisely on the unit circle, then it follows that any nonminimum phase system can be converted to a minimum phase system with the same magnitude frequency response by cascading with an appropriate allpass (AP) system, [21]. The zeros of the NMP system which were located outside the unit circle are moved to the conjugate reciprocal positions within the unit circle of the spectrally equivalent minimum phase (SEMP) system. An example of moving a zero to its reciprocal position within the unit circle is shown in figure (1.8). The term SEMP is used because the operation of

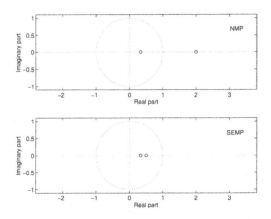

Figure 1.8: Nonminimum Phase System Zero Location and its SEMP System Zero Location

moving the zeros results only in a change in the phase response of the system, the magnitude response remains unaffected. Figure (1.9) shows the effect of moving the zero of figure (1.8) on the phase response.

1.5.3 Phase Blindness of Second-Order Statistics

The loss of phase information using second-order cumulants can be demonstrated by a simple example. Consider the three types of a simple FIR filter with two zeros given by constants a and b. The minimum phase filter, $x_{\text{MP}}[k]$, has both its zeros inside the unit circle, the maximum phase filter, $x_{\text{MXP}}[k]$, has both its zeros outside the unit circle and the nonminimum phase filter, $x_{\text{NMP}}[k]$, has one zero inside, b, and one zero outside, a, the unit circle. The transfer functions of the MP, MXP and NMP filters are given by $H_{\text{MP}}(z)$, $H_{\text{MXP}}(z)$ and $H_{\text{NMP}}(z)$ respectively whilst $x_{\text{MP}}[k]$, $x_{\text{MXP}}[k]$ and $x_{\text{NMP}}[k]$ show the relation between the output, $x[k]$, and the input, $w[k]$, [15].

$$H_{\text{MP}}(z) = (1 - az^{-1})(1 - bz^{-1}) \qquad |a| < 1 \;,\; |b| < 1 \qquad (1.33)$$

$$H_{\text{MXP}}(z) = (1 - az)(1 - bz) \qquad |a| > 1, |b| > 1 \qquad (1.34)$$

$$H_{\text{NMP}}(z) = (1 - az)(a - bz^{-1}) \qquad |a| > 1 \;,\; |b| < 1 \qquad (1.35)$$

$$x_{\text{MP}}[k] = w[k] - (a + b)w[k-1] + abw[k-2] \qquad (1.36)$$

$$x_{\text{MXP}}[k] = w[k] - (a + b)w[k+1] + abw[k+2] \qquad (1.37)$$

$$x_{\text{NMP}}[k] = -aw[k+1] + (1 + ab)w[k] - bw[k-1] \qquad (1.38)$$

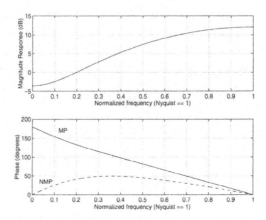

Figure 1.9: Phase Responses of a NMP System and its SEMP System

The second-order cumulants of the output sequences are identical for the MP, MXP and NMP systems

$$
c_{2x}(m) = \begin{cases}
1 + a^2 b^2 + (a + b)^2 & \text{, if } m = 0 \,, \\
-(a + b)(1 + ab) & \text{, if } m = 1 \,, \\
ab & \text{, if } m = 2 \,, \\
0 & \text{, if } m > 2 \,.
\end{cases}
\tag{1.39}
$$

However, the output sequences of the different phase systems do possess different higher-order cumulants. The third-order cumulants of the MP, MXP and NMP systems are given in [16] and repeated in table 1.3. Minimum phase systems can be uniquely identified using second-order cumulants. However, unique identification of a nonminimum phase system requires the use of higher-order cumulants. By comparison, use of second-order cumulants in the identification of a nonminimum phase system will result in the identification of a spectrally equivalent minimum phase system. This SEMP system possesses an identical magnitude distribution to, but a different phase distribution from, the actual nonminimum phase system. The identifiability of both the magnitude and phase of a systems transfer function, $H(z)$, from observations of the output alone depends on the distribution of the input, $w[k]$.

1. If $w[k]$ is Gaussian and $H(z)$ is minimum phase, second-order statistical methods can identify both the magnitude and phase of $H(z)$.

	MP	MXP	NMP
$c_{3x}(0,0)$	$1 - (a+b)^3 + a^3b^3$	$1 - (a+b)^3 + a^3b^3$	$(1+ab)^3 - a^3 - b^3$
$c_{3x}(1,1)$	$-(a+b)^2 - (a+b)a^2b^2$	$-(a+b)^2 - (a+b)a^2b^2$	$-a(1+ab)^2 + (1+ab)b^2$
$c_{3x}(2,2)$	a^2b^2	ab	$-ab^2$
$c_{3x}(1,0)$	$-(a+b) + ab(a+b)^2$	$(a+b)^2 - (a+b)a^2b^2$	$a^2(1+ab) - (1+ab)^2b$
$c_{3x}(2,0)$	ab	a^2b^2	$-a^2b$
$c_{3x}(2,1)$	$-(a+b)ab$	$-(a+b)ab$	$ab(1+ab)$

Table 1.3: Third-order cumulants of MP, MXP and NMP systems

2. If $w[k]$ is Gaussian and $H(z)$ is nonminimum phase, no method can correctly recover the phase of $H(z)$.

3. If $w[k]$ is non-Gaussian and $H(z)$ is nonminimum phase, second-order statistical methods can only correctly identify the magnitude of $H(z)$, and the spectrally equivalent minimum phase system is identified.

4. If $w[k]$ is non-Gaussian and $H(z)$ is nonminimum phase, higher-order statistical methods can estimate both the phase and magnitude of $H(z)$ accurately without any knowledge of the actual distribution of $w[k]$.

Thus, if the input distribution can be assumed to be non-Gaussian, stationary and independent and identically distributed no explicit knowledge of the input is needed in order to identify a system correctly if higher-order statistics are used.

1.6 Cyclostationary Signals and Statistics

The strict sense description of cyclostationarity [7, 19] is a signal which has a joint probability density function which varies periodically with time:

$$\prod_{i=1}^{N} p(x, t_i) = \prod_{i=1}^{N} p(x, t_i + kT) \tag{1.40}$$

where T is the fundamental period of the cyclostationarity and k is an arbitrary integer. Because of this cyclostationary processes have moments and cumulants which vary periodically with time:

$$E\left[\prod_{i=1}^{N} x(t_i)\right] = E\left[\prod_{i=1}^{N} x(t_i + kT)\right] \tag{1.41}$$

where N denotes the order of the statistic. A first-order cyclic-statistical process, $N = 1$, is a periodic signal which may be corrupted with stationary noise:

$$x(t) = a \cos(2\pi f_0 t + \theta) + \eta(t) \tag{1.42}$$

Examples of second-order cyclic-statistical processes [6] include sinusoids amplitude modulated by a random bandlimited signal, and periodic impulses of random noise.

$$x(t) = a(t) \cos(2\pi f_0 t + \theta) + \eta(t) \tag{1.43}$$
$$x(t) = a(t)s(t) + \eta(t) \tag{1.44}$$

where $a(t)$ is a bandlimited random signal and

$$s(t) = \begin{cases} 1 & : & t \pmod{T} < t_m \\ 0 & : & t \pmod{T} > t_m \end{cases}$$

is a periodic rectangular pulse train. Fourth-order cyclostationarity can be observed in quadrature-amplitude modulated signals. In most other cases the first, and second-order cyclic-moments are the most significant, and a signal which exhibits up to second-order cyclostationarity is described as wide-sense cyclostationary.

Since the moments are periodic, they can be expanded into their Fourier series. The Fourier coefficients of the periodic-time varying autocorrelation is termed the cyclic-autocorrelation, and is defined as:

$$R_{xx}^{\alpha}(\tau) = E\left[x(t - \frac{\tau}{2})x(t + \frac{\tau}{2})e^{-j2\pi\alpha t}\right] \tag{1.45}$$

where $\alpha = \frac{k}{T}$ is the kth harmonic of the rotation frequency.

The cyclic-autocorrelation gives an indication of how much energy in the signal is due to cyclostationary components at frequency α. Along the line $\alpha = 0$, lies the stationary autocorrelation of the signal. If a significant amount of energy exists along lines where $\alpha \neq 0$ then this indicates that the signal is cyclostationary. The degree of cyclostationarity (DCS) is defined as [22]:

$$\text{DCS}^{\alpha} = \frac{\int_{-\infty}^{\infty} |R_{xx}^{\alpha}(\tau)|^2 \, d\tau}{\int_{-\infty}^{\infty} |R_x^0(\tau)|^2 \, d\tau} \tag{1.46}$$

A more thorough statistical approach providing a test with a probability of detection can be found in [4].

In the same way that the power spectrum can be obtained by taking the Fourier Transform of the autocorrelation (the Wiener-Khinchin relationship), taking the Fourier Transform of the cyclic-autocorrelation with respect to the lag τ produces the Spectral Correlation Density Function (SCDF) which contains the power spectrum of the signal lying along the $\alpha = 0$ axis.

$$S_{xx}^{\alpha}(f) = \int_{-\infty}^{+\infty} R_{xx}^{\alpha}(\tau)e^{-j2\pi f\tau}d\tau \qquad (1.47)$$

This function gives the correlation between spectral components centred on a frequency f and separated by a frequency shift of α. For the periodic signal given in equation 1.42 the SCDF is:

$$S_{xx}^{\alpha}(f) = \begin{cases} \frac{1}{4}a^2\delta(f - f_0) + \frac{1}{4}a^2\delta(f + f_0) & \text{for} \quad \alpha = 0 \\ \frac{1}{4}a^2 e^{\pm j2\theta}\delta(f) & \text{for} \quad \alpha = \pm 2f_0 \\ 0 & \text{otherwise} \end{cases} \qquad (1.48)$$

With delta functions occurring at the sinusoid frequency along the power spectral axis ($\alpha = 0$) and at zero frequency when $\alpha = 2f_0$. Amplitude modulation has the effect of convolving the power spectrum of the modulated signal with the four delta functions to produce an SCDF which has four bandlimited centred at the positions $(\alpha, f) = \{(0, f_0)(0, -f_0)(2f_0, 0)(-2f_0, 0)\}$. The SCDF for the signal given in equation 1.43 is therefore:

$$S_{xx}^{\alpha}(f) = \begin{cases} \frac{1}{4}S_a(f + f_0) + \frac{1}{4}S_a(f - f_0) & \text{for} \quad \alpha = 0 \\ \frac{1}{4}e^{\pm j2\theta}S_a(f) & \text{for} \quad \alpha = \pm 2f_0 \\ 0 & \text{otherwise} \end{cases} \qquad (1.49)$$

where S_a is the power spectrum of the random signal $a(t)$.

1.7 Estimation of Cyclic-statistics

Cyclostationary signals have the property that under some non-linear transform the signal exhibits periodicity. It is these periodic components which define the cyclostationarity. Therefore a sine wave extraction operation provides a means of determining the moments of a signal related to the fundamental period α,

$$\hat{\mathcal{E}}^{\{\alpha\}}[z(t)] = \sum_{\alpha} \langle z(u)e^{-j2\pi\alpha u}\rangle e^{j2\pi\alpha t} \qquad (1.50)$$

where $\langle . \rangle$ is the time averaging operation [5]. Thus the expected value contains only sinusoidal components which are harmonics of α. This operation can be expressed approximately in a more practical form as the synchronous averaging operation over period $T = 1/\alpha$:

$$\hat{\mathcal{E}}^{\{\alpha\}}\left[z(t)\right] \simeq \frac{1}{N}\sum_{k=0}^{N-1} z(t+kT) \tag{1.51}$$

The cyclic-moment of a periodically time-varying function $z(t)$ can be estimated by taking the discrete Fourier transform of the synchronous average.

$$\mathrm{E}\left[z(t)e^{-j2\pi\alpha t}\right] \simeq \sum_{\alpha}\frac{1}{N}\sum_{k=0}^{N-1} z(t+kT)e^{-j2\pi\alpha t} \tag{1.52}$$

If the cyclic-moment is calculated for the signal $x(t)$, and the averaging period is the rotation period of a rotating mechanical system then the resulting cyclic first-order moment is just the Fourier series expansion of what is more commonly termed the synchronous average. If the cyclic-moment is calculated for the time-varying autocorrelation centred at time t, $z(t) = x(t-\frac{\tau}{2})x(t+\frac{\tau}{2})$, then the resulting second-order moment is defined as the cyclic autocorrelation $R_{xx}^{\alpha}(\tau)$. The use of fractional shifts in the lag variable τ means that if the moment is calculated directly then the frequency resolution is halved. However the time-varying autocorrelation can be calculated without the centring requirement and this can be synchronously averaged. The phase shift introduced can then be compensated for when the moment is transformed from a periodic time-varying function into its Fourier series [8]. For a discrete-time signal this can be achieved using the Discrete Fourier Transform (DFT):

$$R_{xx}^{\alpha}(\tau) = e^{-j\pi\alpha\tau}\mathrm{DFT}_{t\leftrightarrow\alpha}[\mathrm{E}[x(t)x(t+\tau)]] \tag{1.53}$$

The SCDF can subsequently be estimated by taking the discrete Fourier Transform with respect to the time-lag variable τ. This approach to estimating the SCDF is efficient for mechanical vibration signals if the rotation period is known since the values of interest in the SCDF lie at harmonic frequencies of the machine rotation. In some cases the rotation period may not be accurately known, and in other applications of cyclostationary statistics, such as communication signal analysis, the objective may be to identify cyclostationary frequencies in the signal. In such cases the above algorithm is limited since it has a low spectral resolution $\Delta\alpha = \frac{1}{T}$. The size of T can be increased to obtain the required spectral resolution however this is not the most efficient method of determining the SCDF at these resolutions.

More efficient algorithms are based on the time smoothed cyclic cross periodogram:

$$S_{xx}^{\alpha}(f) = \frac{1}{T}\langle X_T(n, f + \alpha/2)X_T^*(n, f - \alpha/2)\rangle \qquad (1.54)$$

The complex demodulates $X_T(n, f)$ can be mathematically expressed as:

$$X_T(n, f) = \sum_{r=-N/2}^{N/2} a(r)x(n - r)e^{-j2\pi f(n-r)} \qquad (1.55)$$

where the number of samples used, N, defines the spectral resolution along the f axis, and $a(r)$ is data tapering window function. These can be efficiently computed using an FFT. The SCDF is then computed by correlating the complex demodulates over the entire time span Δt.

$$S_{xx}^{\alpha}(f) = \sum_{r} X_T(r, f + \alpha/2)X_T^*(n, f - \alpha/2)g(n - r) \qquad (1.56)$$

where $g(n)$ is a data tapering window of length Δt. The efficiency of this process can be improved by two processes. Firstly, decimation can be introduced by computing the complex demodulates only every L samples. This has the effect of reducing the spectral resolution by the decimation factor L. The resolution can be increased however by frequency shifting the product sequence by a small amount.

$$S_{xx}^{\alpha+\epsilon}(f) = \sum_{r} X_T(r, f + \alpha/2)X_T^*(n, f - \alpha/2)g(n - r)e^{-j2\pi\epsilon r} \qquad (1.57)$$

This expresses the correlation operation as a Fourier transform and therefore computing the SCDF for a number of values $(\alpha + \epsilon)$ can be accomplished efficiently using the FFT. This algorithm termed the FFT accumulation method is fully described in [17] along with another efficient high resolution algorithm, the strip spectral correlation algorithm.

1.8 Summary

This chapter has introduced Higher-Order Statistical (HOS) and cyclostationary signal processing and provided definitions of the higher-order moments and cumulants and their statistical properties which are exploited in later chapters to enable blind estimation. Nonminimum phase filters were introduced and the failure of second-order statistics to distinguish between a

NMP filter and its spectrally equivalent minimum phase filter was discussed. This failure is one of the primary reasons for using higher-order cumulants to perform blind equalisation, blind system identification, blind source separation, and many other blind estimation problems.

References

[1] P. O. Amblard and J. M. Brossier. Adaptive estimation of the fourth-order cumulant of a white stochastic process. *Signal Processing*, 42:37–43, 1995.

[2] S. N. Batalama and D. Kazakos. On the robust estimation of the autocorrelation coefficients of stationary sequences. *IEEE Transaction on Signal Processing*, SP-44:2508–2520, 1996.

[3] D. R. Brillinger. *Time series: Data analysis and theory*. Holden-Day Inc., San Francisco, 1981, 1981.

[4] A. V. Dandawate and G. B. Giannakis. Statistical tests for the presence of cyclostationarity. *IEEE Transactions on Signal Processing*, SP-42:2355–2369, Sept 1994.

[5] W. A. Gardner. *Statistical Spectral Analysis: A Non-probabilistic Theory*. Prentice-Hall, Englewood Cliffs, N.J., 1987.

[6] W. A. Gardner. Exploitation of spectral redundancy in cyclostationary signals. *IEEE Signal Processing Magazine*, 8(2):14–36, April 1991.

[7] W. A. Gardner and C. M. Spooner. The cumulant theory of cyclostationary time-series, part i: Foundation. *IEEE Transactions on Signal Processing*, SP-42:3387–3408, Dec 1994.

[8] S. Haykin. *Adaptive Filter Theory*, chapter 3. Prentice Hall, 3rd edition, 1996.

[9] R. Leipnik. The lognormal distribution and strong non-uniqueness of the moment problem. *Theory Prob. Appl.*, 26:850–852, 1981.

[10] G. C. W. Leung and D. Hatzinakos. Implementation aspects of various higher-order statistics estimators. *J. Franklin Inst.*, 333B:349–367, 1996.

[11] D. Mämpel, A. K. Nandi, and K. Schelhorn. Unified approach to trimmed mean estimation and its application to bispectrum of eeg signals. *J. Franklin Inst.*, 333B:369–383, 1996.

[12] S. L. Marple Jr. *Digital Spectral Analysis with Applications*. Prentice Hall, Englewood Cliffs, New Jersey, 1987.

[13] The MathWorks Inc. *Matlab Reference Guide*, 1995.

[14] A. K. Nandi and D. Mämpel. Development of an adaptive generalised trimmed mean estimator to compute third-order cumulants. *Signal Processing*, 57:271–282, 1997.

[15] C. L. Nikias. Higher-order spectral analysis. In S. S. Haykin, editor, *Advances in Spectrum Analysis and Array Processing*, volume I, Englewood Cliffs, New Jersey, 1991. Prentice Hall.

[16] C. L. Nikias and A. P. Petropulu. *Higher-Order Spectra Analysis: A Nonlinear Signal Processing Approach*. Prentice Hall, Englewood Cliffs, New Jersey, 1993.

[17] R. S. Roberts, W. A. Brown, and H. H. Loomis. Computationally efficient algorithms for cyclic spectral analysis. *IEEE Signal Processing Magazine*, 8(2):38–49, April 1991.

[18] O. Shalvi and E. Weinstein. New criteria for blind deconvolution of nonminimum phase systems (channels). *IEEE Transactions on Information Theory*, IT-36:312–321, 1990.

[19] C. M. Spooner and W. A. Gardner. The cumulant theory of cyclostationary time-series, part ii: Development and applications. *IEEE Transactions on Signal Processing*, SP-42:3409–3429, Dec 1994.

[20] A. Stuart and J. K. Ord. *Kendall's Advanced Theory of Statistics*. Charles Griffin and Company, London, 5 edition, 1987.

[21] C. W. Therrien. *Discrete Random Signals and Statistical Signal Processing*. Prentice Hall, Englewood Cliffs, New Jersey, 1992.

[22] G. D. Zivanovic and W. A. Gardner. Degrees of cyclostationarity and their application to signal detection and estimation. *Signal Processing*, 22:287–297, Mar 1991.

2 BLIND SIGNAL EQUALISATION

S N Anfinsen, F Herrmann and A K Nandi

Contents

2.1	Introduction		29
	2.1.1	Applications	29
	2.1.2	Signal Model	30
	2.1.3	Equalisation Criterion	33
	2.1.4	Conditional Mean Estimator	34
	2.1.5	Inverse Modelling of a Nonminimum Phase System	36
	2.1.6	Digital Communications Context	38
2.2	Gradient Descent Algorithms		40
2.3	Blind Equalisation Algorithms		42
	2.3.1	Gradient Calculation	43
	2.3.2	Non-convexity of the Cost Function	44
	2.3.3	Sato Algorithm	44
	2.3.4	Constant Modulus Algorithm	45
	2.3.5	Benveniste-Goursat Algorithm	47
	2.3.6	Stop-and-Go Algorithm	48
	2.3.7	Simulation Environment	49
	2.3.8	Simulations with Bussgang Algorithms	50
2.4	Algorithms Based on Explicit HOS		55
	2.4.1	Tricepstrum Equalisation Algorithm	56
		Definition of Polycepstra	56
		Decomposition of Non-minimum Phase Channel	57
		Channel Estimation	58
	2.4.2	Maximum Kurtosis Criterion	59

2.4.3   Super-Exponential Algorithm   60

2.4.4   Eigenvector Approach   62

2.4.5   Adaption by First-Order Stochastic Approximation. . .   64

2.4.6   Simulations with Explicit HOS-Based Algorithms   66

2.5   Equalisation with Multiple Channels   67

2.5.1   Fractionally Spaced Equalisation   67

2.5.2   Equalisation with Multisensor Arrays   69

2.5.3   Multichannel Signal Model   70

2.5.4   Single Input Single Output Model   73

2.5.5   Single Input Multiple Output Model   76

2.5.6   Full Rank Condition   78

2.5.7   Extending T-Spaced Algorithms to Fractionally Spaced
        Equalisation   81

2.5.8   Simulations with FSE Algorithms   82

2.6   Algorithms Based on Cyclostationary Statistics   84

2.6.1   Cyclostationarity of Modulated Input   86

2.6.2   Spectral Diversity   87

2.6.3   Overview of CS-Based Algorithms   88

2.6.4   Zero Forcing Algorithm   88

2.6.5   Subspace Decomposition   90

2.7   General Convergence Considerations   93

2.7.1   Gradient Descent Algorithms   93

2.7.2   Algorithm and Channel Dependent Equilibria   93

2.8   Discussion   95

References   97

2.1 Introduction

The objective of equalisation is to design a system that optimally removes the distortion that an unknown channel induces on the transmitted signal. This is in effect inverse system modelling, an architecture that is well-known in adaptive filtering theory. The cascade of channel and equaliser should constitute an identity operation, with the exception of a time delay and linear phase shift being allowed.

In non-blind equalisation, the equalisation filter is chosen so that the equaliser output matches the observable input signal. Blind equalisation, on the other hand, is performed without access to the original input. It can also be termed unsupervised or self-recovering equalisation, since we do not have a known training sequence, a target or desired signal in terms of adaptive filtering. This makes the problem significantly more complex. Blind equalisation is the same problem as blind deconvolution. The aim of both is to recover the unobservable excitation signal given the response of an unknown system.

The approach used in blind equalisation is to equalise statistics of the output signal with statistics of the input signal. Benveniste et al. have shown [3] that equalisation is obtained if the input and output signal has the same probability density function (pdf). Hence, the distribution is all preliminary information required about the input. Different techniques have been developed to solve this problem, and some of the most important ones will be discussed in this chapter.

2.1.1 Applications

The nature of the problem implies that non-blind methods in most situation yield better results than blind solutions, considering convergence speed and equalisation quality. Also, the blind approach must clearly have a higher computational cost. Non-blind equalisation is widely used in digital wireless communication systems like GSM (Global System for Mobile Communication).

Mobile phone signals are subject to severe distortion, due to reflection and diffraction of the radio wave carrier. To combat the effects of the multipath environment, GSM relies on periodical retransmission of a known bit-sequence. The GSM receiver then estimates the impulse response of the medium from the received signal, and models an equaliser that unravels the effect of the distortion.

But there are other aspects that make blind equalisation attractive. It is sometimes desirable to start up the receiver of a communications system without resorting to a training sequence. The first research efforts on blind equalisation emerged from problems of multipoint data networks. Terminal equipment of such networks need equalisation to be able to read data and system messages. But terminals might be powered on after initial network synchronisation. Since lines are shared, simultaneous access may cause collisions and interrupted messages. Training sequences risk being interrupted, as well as adding excessive load to the network. Thus, self-recovering equalisation is highly appropriate.

Blind equalisation has advantages in systems where constant-rate retransmission of a known sequence is too costly, since such practice necessarily reduces channel capacity. In GSM, the training sequence accounts for a 22% overhead in transmitted data. Furthermore, in cases of severe distortion, the training sequence might be too short to give a good estimate of the inverse channel. A blind equaliser utilises the transmitted signal in full length, rather than dedicated sequences. It may therefore be able to track changes in a time-varying and non-predictable channel, if not limited by convergence speed.

Consequently, blind equalisers are important parts of many high-speed communication systems. They are used to compensate for the described effects of multipath propagation, as found in digital wireless communication and underwater acoustics. As mentioned, they are also used to restore channels in telephone and multipoint data networks that may experience considerable echoing and distortion.

Another application field is reflection seismology, where the solution is commonly referred to as blind deconvolution. Seismic exploration is performed by generating an acoustic wave field that is reflected by geological layers with different acoustic impedance. Blind deconvolution is used to remove the source waveform and other undesirable influences from the seismograms. Yet another application is found in image processing, where blind deconvolution is used for purposes of deblurring and image restoration. For instance, blind deconvolution techniques are of great practical importance in astronomical imaging.

2.1.2 Signal Model

Figure 2.1 shows a basic configuration of a blind equaliser incorporated in a communications system. As a foundation for subsequent analysis we define a discrete signal model where all random sequences and filters are generally complex.

An unobservable input sequence $x(n)$ is passed through an unknown channel $h(n)$. We assume that $x(n)$ is a sequence of independently and identically

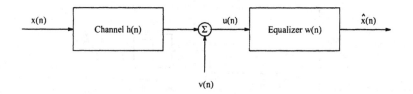

Figure 2.1: Block diagram of blind equaliser.

distributed (i.i.d.) symbols. $h(n)$ denotes the composite channel impulse response, including transmitter and receiver filters as well as the physical medium. The channel $h(n)$ is linear and stable. Here we will assume that it is time-invariant. In practice, the equaliser will be able to track time-varying channels in an on-line application, if the change is slow compared to the time it takes to train the equaliser (piece-wise time-invariant channel).

Most physical channels can be approximated by a discrete time and finite length impulse response. Accurate modelling is feasible as long as the tap spacing of $h(n)$ is less than the inverse bandwidth of the transmitted signal [1]. Still, the length of $h(n)$ is theoretically infinite. The channel is possibly nonminimum phase, which means that it may have poles both inside and outside the unit circle of the z-plane. The inverse of the channel exists, but is possibly non-causal (due to maximum phase).

The input data experience intersymbol interference (ISI) because of the non-ideal characteristics of the channel. This means that the channel output symbol will be a linear combination of the present input and previous input symbols. The transmitted sequence is also corrupted by additive noise $v(n)$, which we assume is white and Gaussian. The noise term covers all additional interferences of the system. The transmitted sequence arrives at the receiver as the sequence $u(n)$

$$u(n) = h(n) * x(n) + v(n)$$
$$= \sum_{k=-\infty}^{+\infty} h(k)x(n-k) + v(n) \,. \qquad (2.1)$$

The objective is to deconvolve the receiver input u(n) to retrieve the channel input. This is done by passing $u(n)$ through a blind equaliser with finite impulse response (FIR) $w(n)$ of length $L+1$. The output of the blind equaliser

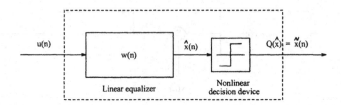

Figure 2.2: Receiver configuration with equaliser and decision circuit.

is denoted $\hat{x}(n)$

$$\hat{x}(n) = w(n) * u(n)$$
$$= \sum_{k=0}^{L} w(k)u(n-k) \, . \tag{2.2}$$

An on-line application of a blind equaliser will involve two phases. Initially, we assume no knowledge at all about the channel. This is the start-up period where an algorithm is used to obtain an initial estimate of the inverse system that equalises the distorting channel. The start-up condition for the equaliser is sometimes referred to as the closed-eye condition.

When acceptable performance, for instance a pre-determined error-level is reached, the equaliser is switched into the decision-directed mode. This is the stage when equalisation is sufficiently good to make correct decisions about the input data. Alternatively, we call this the open eye condition.

If we assume that the data source is discrete, the equaliser output will now be passed through a decision device or quantiser $Q(\hat{x})$. The whole receiver configuration is shown in figure 2.2. The quantiser is a maximum likelihood estimator, which reduces the mean squared error (MSE) of the equaliser output. It decides which of the symbols in the finite alphabet of the discrete data source is closest to $\hat{x}(n)$. The result is an estimate $\tilde{x}(n)$ that equals $x(n)$ when the decision error rate is zero (eye is open)

$$\tilde{x}(n) = Q\big(\hat{x}(n)\big)$$
$$= Q\big(h(n) * w(n) * x(n)\big) \, . \tag{2.3}$$

Channel input $x(n)$ and noise $v(n)$ are sequences of zero-mean, independently and identically distributed (i.i.d.) symbols. The only constraint on the input pdf is that it must be non-Gaussian. This is because the equalisation will rely on higher-order statistics. These are shown to be identically zero, and therefore not useful, for Gaussian signals. This model will be used as a basis for the approaches described in subsequent sections.

2.1.3 Equalisation Criterion

For perfect equalisation, we want the equaliser to remove all influences of the distorting channel. The cascade of channel and equaliser is denoted $s(n)$. We want the overall impulse response of the cascade to be an identity operation. But we allow the equaliser to impose gain factor, a constant delay and a linear phase shift on the transmitted signal. It effectively means that

$$
\begin{aligned}
s(n) &= h(n) * w(n) \\
&= \delta(n-k)\, c\, e^{j\theta}
\end{aligned}
\tag{2.4}
$$

where k is the delay, c is the gain factor and θ the linear phase shift. This is known as the equalisation criterion in the s-domain. In other words, we want the energy of the overall impulse response to be confined to one tap

$$
s(n) = s(k)\delta(n-k) \ .
\tag{2.5}
$$

However, this criterion is not of any direct practical value. We cannot determine $w(n)$ from Eq. (2.4), since $h(n)$ is unknown. But it can help us in the search for realisable algorithms in the w-domain. Further, it provides a numerical measure of ISI that can be useful in simulations. This is given by

$$
\mathrm{ISI} = \frac{\sum_k |s[k]|^2 - |s(n_{\max})|^2}{|s(n_{\max})|^2}
\tag{2.6}
$$

where $s(n_{\max})$ is the component of the overall impulse response with greatest magnitude. It is evident from the s-domain criterion that perfect equalisation gives zero ISI. Another common performance measure for equalisation algorithms is MSE, defined for a sequence of N symbols as

$$
\mathrm{MSE}(N) = \frac{1}{N} \sum_{n=1}^{N} |\hat{x}(n) - x(n)|^2 \ .
\tag{2.7}
$$

Performance can also be assessed by the symbol error rate (SER), which is simply the rate of decision errors made by the decision circuit.

As declared in the signal model, this presentation is restricted to FIR blind equalisers. Due to this fact, perfect equalisation of the described single input-single output (SISO) system is theoretically limited. The convolution of a truncated length equaliser with a generally infinite channel impulse response cannot produce the desired delta function, or equivalently guarantee complete removal of intersymbol interference.

2.1.4   Conditional Mean Estimator

Let $\hat{w}(n)$ be a finite length estimate of the possibly infinite perfect equaliser. Hence, the output of the estimated equaliser is

$$\hat{x}(n) = \sum_k \hat{w}(k)u(n-k) \, .$$

We can rewrite this as

$$\hat{x}(n) = \sum_k w(k)u(n-k) + \sum_k \big(\hat{w}(k) - w(k)\big)u(n-k) \qquad (2.8)$$

and define the convolutional error

$$\eta(n) = \sum_k \big(\hat{w}(k) - w(k)\big)u(n-k) \, . \qquad (2.9)$$

Thus, the equaliser output can be written as a sum of the true input and a convolutional error

$$\hat{x}(n) = x(n) + \eta(n) \, . \qquad (2.10)$$

The convolutional error $\eta(n)$ represents the residual intersymbol interference induced by channel distortion. The additive noise $v(n)$ can, and will in the derivation of many algorithms, be disregarded, since it is negligible compared to the initial convolutional error and will only have a considerable effect after much ISI is removed. The matter would be different if the channel was exposed to impulse noise. This is a special case which will not be treated in this text.

If we define the residual impulse response due to the nonideal channel characteristics as

$$\psi(n) = h(n) * \big(\hat{w}(n) - w(n)\big) \qquad (2.11)$$

then we have from Eq. (2.9) that

$$\eta(n) = \sum_k \psi(k)x(n-k) \, . \qquad (2.12)$$

From certain assumptions about the residual impulse response, it can be deduced that the random sequence $\eta(n)$ is approximately:

1. zero mean,

2. Gaussian and

3. independent of the input sequence [21].

This provides the basis for adaption to recover the input sequence. The estimation of $x(n)$ from Eq. (2.10) is a classical problem treated in the literature [49]. We can derive a conditional estimate of the unobservable desired signal, denoted $\hat{d}(n)$, given the observation of the equaliser output $\hat{x}(n)$. Suppressing the time variable of the random sequences, we have

$$\hat{d}(n) = \mathrm{E}[x(n)|\hat{x}(n)]$$
$$= \int_{-\infty}^{+\infty} x(n) f_X\big(x(n)|\hat{x}(n)\big)\, dx \qquad (2.13)$$

where $f_X\big(x(n)|\hat{x}(n)\big)$ is the conditional probability density function (pdf) of $x(n)$, given $\hat{x}(n)$. For the instant, we shall suppress the time index for convenience. Bayes' theorem states

$$f_X(x|\hat{x}) = \frac{f_{\hat{X}}(\hat{x}|x)f_X(x)}{f_{\hat{X}}(\hat{x})} \qquad (2.14)$$

with the respective pdfs of x and $\hat{x}$ denoted $f_X(x)$ and $f_{\hat{X}}(\hat{x})$. $f_{\hat{X}}(\hat{x}|x)$ is the conditional pdf of $\hat{x}$, given x. Substituted into the conditional estimator, this becomes

$$\hat{d} = \frac{1}{f_{\hat{X}}(\hat{x})} \int_{-\infty}^{+\infty} x\, f_{\hat{X}}(\hat{x}|x)f_X(x)\, dx \ . \qquad (2.15)$$

A transformation of random variables with help of Eq. (2.10) yields

$$f_{\hat{X}}(\hat{x}|x) = f_\eta(\hat{x} - x) \ . \qquad (2.16)$$

With the assumptions of $v(n)$ as a zero-mean and Gaussian random sequence, this expression is inserted into Eq. (2.15). The result is a Bayes estimator that can be evaluated for any specific input distribution

$$\hat{d} = \frac{1}{f_{\hat{X}}(\hat{x})} \int_{-\infty}^{+\infty} x\, f_\eta(\hat{x} - x)f_X(x)\, dx \ . \qquad (2.17)$$

It can be shown that this is a minimum mean squared error estimator. Still, we miss one parameter for a complete definition of the Gaussian pdf $f_\eta(\hat{x} - x)$, namely the variance σ_η^2. Also, even though the estimator is optimal in a mean squared error sense, it relies on an approximation of $f_\eta(\eta)$. The assumptions made in the deduction of $f_\eta(\eta)$ are not trivial. The conditional mean estimator is therefore only a suboptimal solution.

The task of removing intersymbol interference, or equivalently the convolution error, requires nonlinear methods. The placement of the nonlinearity within the blind equaliser structure varies in the different techniques which have evolved. The examination in the next section will make clear how we can approach the problem from a statistical view.

2.1.5   Inverse Modelling of a Nonminimum Phase System

We have assumed that the channel has unknown, but stationary transfer function. It is also a possibly nonminimum phase system, which means that it may have zeros outside the unit circle on the z-plane. Assume that $x(n)$ is at least wide sense stationary (wss). The excitation signal is also (higher order) spectrally white (from the i.i.d. assumption). For such input it is well known that the magnitude response of an unknown system can be identified from second-order statistics like the autocorrelation function or the power spectrum.

For a non-minimum phase system, there is a one-to-one relationship between the magnitude response and the phase response. For non-minimum phase systems, on the other hand, phase information is only preserved in statistics of order higher than two. This enforces equalisation algorithms based on higher-order statistics (HOS), implicitly or explicitly. Common for all these approaches is that they exhibit slow convergence rate [21]. This is because the number of samples required to obtain adequate empirical estimates rises almost exponentially with the order of the higher-order statistics [4].

However, phase information about the channel can be extracted from second-order statistics, but this requires other properties of the excitation signal. This is feasible for cyclostationary, or periodically correlated signals, as discovered by Gardner [15]. These results has motivated research on methods using cyclostationary statistics (CS). The bulk of blind equalisation algorithms can thus be divided into three main categories.

- *Gradient Descent Algorithms*, that emulate the structure of conventional adaptive non-blind equalisers. The training sequence in the non-blind inverse modelling architecture is replaced by a nonlinear estimate of the channel input. The nonlinearity is designed to minimise a cost function that is implicitly based on higher-order statistics (HOS). An adaptive equalisation filter is updated through a gradient descent algorithm. Bussgang algorithms are the most prominent members of this class.

- *Algorithms Based on Explicit Higher-Order Statistics*, that use higher-order cumulants and polyspectra. Expressions can be found that relate the equaliser solution directly to such higher-order statistics. Practical realisations rely on estimation of empirical statistics. This requires nonlinear computations involving equaliser input, and possibly output, depending on the algorithm.

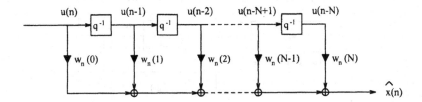

Figure 2.3: Implementation of transversal FIR filter.

- *Algorithms Based on Cyclostationary Statistics.* Cyclostationarity is imposed on the equaliser input signal by oversampling of the received signal with respect to the symbol rate (baud rate), or employing a multi-sensor array at the receiver. The resulting spectral diversity enables modelling of the inverse system. Channel equalisation can be achieved by a zero-forcing algorithm.

The latter family introduces the division between symbol rate spaced and fractionally spaced equalisation algorithms. Fractional spacing of the equaliser input with respect to the baud rate is a concept that can also be employed for algorithms of the first two categories. Motivation and implementation of fractionally spaced equalisation is addressed in subsequent sections.

All the mentioned groups of algorithms commonly realises the equaliser as a transversal filter, as shown in figure 2.3. This is the most common implementation of an FIR filter. Other filter structures have been investigated for special applications, although not with the same broad interest. An IIR equaliser with recursive structure may offer attractive features, after inherent problems like stability and complexity are addressed. These benefits include increased ability to equalise channels with sharp resonances (zeros close to the unit circle) and reduced number of filter taps, as well as the theoretical ability of perfect equalisation of SISO systems. Another possibility is to employ a nonlinearity within the filter itself. Such approaches has been attempted with nonlinear filters like Volterra models and neural network architectures. The decision feedback equaliser (DFE) is another nonlinear structure that is well known from non-blind equalisation. In the DFE, the input to the decision circuit consists of both filtered received signal and a filtered version of the decision circuit output, as shown in figure 2.4. The main disadvantage is that the the feedback path propagates decision errors, which creates problems in start-up mode. However, a study of these structures is outside the scope of our text, as we shall confine ourselves to linear equalisers.

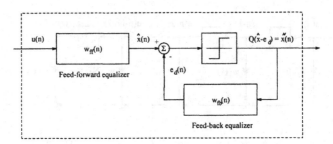

Figure 2.4: Structure of decision feedback equaliser.

2.1.6 Digital Communications Context

In the rest of this chapter, the blind equalisation problem will be considered in a digital communications context. This includes the assumption that the input data to the communications channel is generated by a discrete memoryless source (DMS). Examples of digital transmission systems include mobile phones, microwave radio, troposcatter radio, digital TV, cable TV, voiceband modems and data networks.

Before it is transmitted over the communications link, a sequence of raw binary data may undergo several operations like error coding, channel coding, encryption and modulation. In the modulator, the data sequence is commonly transformed into a representation with multilevel symbols to increase channel capacity [35, 40]. This means that the alphabet used to represent data has M (more than two) symbols. It is also customary in modern modems to use two dimensional modulation schemes. Symbols have amplitude as well as phase, and are visualised in complex space with real (in-phase) and imaginary (quadrature-phase) parts.

Performance of the algorithms presented in the sequel will be illustrated experimentally. In the simulations we will use data that are random-generated from modulation schemes in common use. We have chosen QAM-16 and V.29 as examples.

Quadrature amplitude modulation (QAM) consists of two independently amplitude-modulated carriers in quadrature. Thus, it can be considered a logical extension of quadrature phase shift keying (QPSK). QPSK uses an alphabet of 4 symbols with equal amplitude, but different phase. QAM-M combines one or more ($M/4$) sets of QPSK constellations with different amplitude. Thus, it can be viewed as a combination of amplitude shift keying and phase shift keying, giving rise to the alternative name phase amplitude modulation (PAM). It can also be seen as amplitude shift keying in two

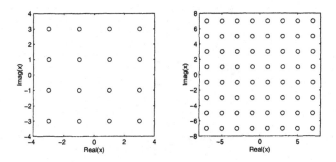

Figure 2.5: Signal space diagrams of QAM-16 and QAM-64 source.

dimensions, termed quadrature amplitude shift keying (QASK). We will assume that QAM signals are modulated with amplitude values

$$\{\cdots, \; -5, \; -3, \; -1, \; 1, \; 3, \; 5, \; \cdots\}$$

although other choices are possible. Figure 2.5 shows the signal spaces of the QAM-16 and QAM-64 constellations. The CCITT V.29 standard is a PAM constellation used in many commercial modems. It can also be seen as amplitude shift keying in two dimensions, but real and imaginary part of the symbols are not independent. This is an important difference between V.29 and QAM-M. A V.29 source admits an alphabet of 16 equi-probable values to be transmitted:

$$\begin{aligned}
\{1+\jmath, \quad & 1-\jmath, \quad -1+\jmath, \quad -1-\jmath, \\
3, \quad & -3, \quad -3\jmath, \quad 3\jmath, \\
3+3\jmath, \quad & 3-3\jmath, \quad -3+3\jmath, \quad -3-3\jmath, \\
5, \quad & -5, \quad 5\jmath, \quad -5\jmath\}
\end{aligned}$$

A signal space diagram of the V.29 transmission values is shown in figure 2.6. Finally, some remarks about the importance of good adaptive equaliser design in digital transmission systems [23]:

- Adaptive equalisers are crucial for successful employment of multilevel modulation schemes. Such schemes are used to obtain maximum transmission rate and optimal use of allocated bandwidth in systems with restricted resources. But they also make signal recovery increasingly difficult. The capability of the equaliser is the limitation.

- The equaliser is the most important part of the demodulator, and also the most computationally intensive. In transmission with QAM-64

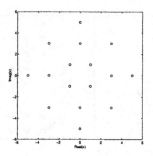

Figure 2.6: Signal space diagram of V.29 source.

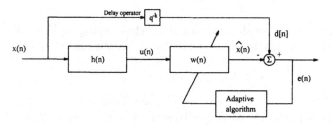

Figure 2.7: Block diagram of adaptive non-blind equaliser.

data, equaliser operations typically account for 80% or more of the multiply-and-accumulate (MAC) cycles in the demodulator.

2.2   Gradient Descent Algorithms

Gradient descent blind equalisation algorithms use an iterative procedure to reduce the convolutional error. The idea is that if we can find a cost function $J(n)$ that characterises intersymbol interference, then minimising this measure with respect to the equaliser parameters will also reduce the convolutional error. This coupling has been obtained by cost functions that are implicitly based on higher-order statistics.

In the iterative equalisation scheme, a refined equaliser estimate is computed at the arrival of each new symbol. The algorithm "learns" by adapting filter taps that minimises the cost function. Minimisation is performed with the gradient descent algorithm, which is a widely used parameter estimation technique [21]. The algorithm searches for an optimal filter tap setting by moving in the direction of the negative gradient $-\nabla_{\mathbf{w}} J(n)$ over the surface of

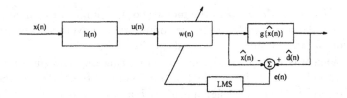

Figure 2.8: Block diagram of adaptive blind equaliser updated with gradient descent algorithm.

the cost function in the equaliser filter tap space. Thus, the update equation is given by

$$w_k(n) = w_k(n-1) + \mu\left(-\nabla_{w_k} J(n)\right)^*, \qquad \text{for} \qquad k = 0, \ldots, L \qquad (2.18)$$

where $w_k(n)$ denotes the kth filter tap at the arrival of equaliser input $u(n)$. The parameter μ determines the step-size of the gradient search and influences therefore the stability and convergence behaviour of the algorithm.

Non-blind adaptive equalisation algorithms seek to minimise a cost function like the mean square error (MSE)

$$J_{\mathrm{MSE}}(n) = \mathrm{E}\left[|e_t(n)|^2\right] \qquad (2.19)$$

where $e_t(n)$ is the true estimation error of the adaptive filter, defined by

$$e_t(n) = \hat{x}(n) - d(n) . \qquad (2.20)$$

The structure of a non-blind equaliser is shown in figure 2.7. The desired signal, denoted $d(n)$, is a possibly delayed version of the channel input. The computationally much simpler least mean square (LMS) algorithm is based on minimising the instantaneous error power (the expectation operation from Eq. (2.19) is dropped here)

$$J_{\mathrm{LMS}}(n) = |e_t(n)|^2. \qquad (2.21)$$

The LMS algorithm is well suited for on-line applications, but reduced complexity is obtained on the expense of low convergence speed.

The blind equalisation approach presented in the sequel can be seen as a logical extension of LMS-type non-blind equalisation techniques. The challenge of the blind configuration is to find a substitute for the absent training sequence d(n). Since the system input is unobservable, a target for adaption has to be estimated from the available data. A solution is to employ a non-linearity at the equaliser output, thus obtaining an estimate based on $\hat{x}(n)$

to $\hat{x}(n - M)$, where M denotes the memory of the nonlinear estimator. The result is an algorithm with very little computational overhead as compared to the comparatively simple non-blind architecture.

This idea was first presented in the pioneering paper of Sato [36], that initiated much research in the field of blind equalisation. Here it was suggested to replace the desired signal $d(n)$ with an estimate of $x(n)$ based only on the present equaliser output $\hat{x}(n)$. The proposed estimator is a memoryless nonlinearity, which we denote $g(\hat{x}(n))$. Sato's algorithm was the first of the Bussgang algorithms, which got their name because they employ memoryless nonlinearities whose output assume Bussgang statistics [5] when they convergence in the mean value. The analogy between non-blind and blind gradient descent techniques can be seen from figure 2.8.

2.3   Blind Equalisation Algorithms

Sato introduced the idea of employing a memoryless nonlinear estimator $g(\hat{x}(n))$ to produce a substitute for the training signal. The output of the nonlinearity is an estimate $\hat{d}(n)$ of the desired signal, which is used to compute an error signal $e(n)$.

$$
\begin{aligned}
e(n) &= \hat{x}(n) - d(n) \\
     &= \hat{x}(n) - g(\hat{x}(n)) \ .
\end{aligned}
\tag{2.22}
$$

For complex baseband channels, as we have generally assumed our systems to be, real and imaginary part of the signal are processed separately by the nonlinearity. The resulting error signal is

$$
e(n) = g(\hat{x}_{\mathrm{re}}(n)) + \jmath\, g(\hat{x}_{\mathrm{im}}(n))\hat{x}(n) \ .
\tag{2.23}
$$

where $x_{\mathrm{re}}(n)$ and $x_{\mathrm{im}}(n)$ denote real and imaginary part. The error signal is then used in the gradient descent algorithm. At each iteration, the filter parameters are changed in the direction of the negative gradient. The gradient with respect to the equaliser filter $\mathbf{w}$ is defined as

$$
\nabla_{\mathbf{w}} J(n) = \left[\frac{\partial J(n)}{\partial w_0(n)} \quad \cdots \quad \frac{\partial J(n)}{\partial w_L(n)} \right]^T
\tag{2.24}
$$

where $J(n)$ is the cost function that we want to minimise. The choice of cost function is a crucial part of all adaptive algorithms. It will determine the nonlinear estimator to be used, as $g(\hat{x}(n))$ results from the minimising of the cost function with respect to the equaliser filter taps.

2.3.1   Gradient Calculation

The adaptive algorithm is derived for a transversal filter vector

$$\mathbf{w}(n) = \begin{bmatrix} w_0(n) & \dots & w_L(n) \end{bmatrix}^T$$

with $L + 1$ filter taps corresponding to the finite impulse response of the equaliser. The cost function is given explicitly in $y(n)$. Minimisation with respect to $\mathbf{w}(n)$ is performed through a complex differentiation, which is rather simple since equaliser input and output are related through the convolution

$$\hat{x}(n) = \sum_{k=0}^{L} w_k(n)\, u(n - k) \,. \tag{2.25}$$

The gradient calculation for $J(n)$ can be written as

$$\begin{aligned}
\nabla_{w_k} J(n) &= \frac{\partial J(n)}{\partial w_k(n)} \\
&= \frac{\partial J(n)}{\partial \hat{x}(n)} \frac{\partial \hat{x}(n)}{\partial w_k(n)} \\
&= \frac{\partial J(n)}{\partial \hat{x}(n)}\, u(n - k) \,.
\end{aligned} \tag{2.26}$$

For the LMS cost function, this leads to

$$\nabla_{w_k} J_{\mathrm{LMS}}(n) = \underbrace{\big(x(n) - d(n)\big)^*}_{e^*(n)} u(n - k)$$

By comparing the latter equation with Eq. (2.26) it becomes obviously to define the error signal

$$e^*(n) = \frac{\partial J(n)}{\partial \hat{x}(n)} \,. \tag{2.27}$$

By analogy, the nonlinear estimator is then determined from Eq. (2.22)

$$g\big(\hat{x}(n)\big) = \hat{x}(n) - \left(\frac{\partial J(n)}{\partial \hat{x}(n)}\right)^* \,. \tag{2.28}$$

Different cost functions have been suggested, leading to various nonlinearities. Some of these will be considered in the following. The common feature is that they make implicit use of higher-order statistics.

2.3.2   Non-convexity of the Cost Function

In a compact filter vector notation, the update equation for the taps of the FIR filter looks quite similar to the LMS algorithm for the non-blind case

$$\mathbf{w}(n) = \mathbf{w}(n-1) - \mu \, \mathbf{u}^*(n) \, e_t(n) \qquad (2.29)$$

where $\mathbf{w}(n)$ is the filter vector at the arrival of equaliser input $u(n)$ and the input vector $\mathbf{u}(n) = \begin{bmatrix} u(n) & \cdots & u(n-N) \end{bmatrix}^T$. The major difference between the non-blind and the blind case is that non-blind gradient descent algorithms bases on a quadratic cost function and the gradient search on the convex error surface in the equaliser vector space will therefore lead to global convergence and an optimal filter. The unique minimum is approached by the gradient descent algorithm under the condition that the step size parameter is small enough to provide convergence. In contrast, the blind cost function $J(n)$ is a nonlinear function that may exhibit local minima. It does not possess the desired unimodality. As a consequence, adaptive blind equalisation algorithms may be trapped in local minima, producing sub-optimal solutions. This is known as ill-convergence. We will next look at some special-case blind algorithms.

2.3.3   Sato Algorithm

In Sato's paper, blind equalisation was considered in the context of amplitude modulated data transmission only. The following cost function was proposed without any theoretical justification [36].

$$J_{\mathrm{S}}(n) = \mathrm{E}\left[\frac{1}{2}\hat{x}^2(n) - \gamma \, |\hat{x}(n)| \right] \, . \qquad (2.30)$$

where the Sato parameter γ is a constant defined as

$$\gamma = \frac{\mathrm{E}[x^2(n)]}{\mathrm{E}[|x(n)|]} \, . \qquad (2.31)$$

In the literature, Sato's cost function is commonly reformulated as

$$J_{\mathrm{S}}(n) = \mathrm{E}\left[\gamma \mathrm{sign}(\hat{x}(n)) - \hat{x}(n) \right)^2 \right] \qquad (2.32)$$

where $\mathrm{sign}(\cdot)$ is the signum function. Note that for a symmetric constellation of complex data symbols, $\mathrm{E}[x^2(n)]$ is zero. To account for phase amplitude modulated data transmission, the Sato parameter must be redefined. With a two-dimensional carrier, we can process the real and imaginary part of the signal separately and let

$$\gamma_{\mathrm{re/im}} = \frac{\mathrm{E}\left[\left|\left(x_{\mathrm{re/im}}(n)\right)\right|^2\right]}{\mathrm{E}\left[\left|\left(x_{\mathrm{re/im}}(n)\right)\right|\right]} \tag{2.33}$$

since, if the PAM symbol constellation is symmetric, which is normally the case, then $\gamma_{\mathrm{re}} = \gamma_{\mathrm{im}} = \gamma$. Splitting the real and imaginary part of the signal is not unproblematic, though. In modulation schemes like V.29, the carriers are correlated. Thus, treating them separately may deteriorate the behaviour of the algorithm. The error signal is derived from Eq. (2.30) as

$$e_{\mathrm{S}}(n) = \hat{x}(n) - \gamma\mathrm{sign}(\hat{x}(n)) \,. \tag{2.34}$$

The nonlinearity hence becomes with Eq. (2.28)

$$g_{\mathrm{S}}(\hat{x}(n)) = \gamma\mathrm{sign}(\hat{x}(n)) \,. \tag{2.35}$$

Sato's nonlinearity treats both real and imaginary part of its input as a binary signal, and produces a scaled estimate of the sign bit.

2.3.4 Constant Modulus Algorithm

An improved family of cost functions was suggested by Godard [19]. These cost functions involve only the magnitude of $\hat{x}(n)$. The effect is to characterise ISI independently of the phase of the equaliser output. Godard's general cost function, termed the dispersion of order p, is given by

$$J_{\mathrm{G}}(n) = \frac{1}{p}\mathrm{E}\left[(|\hat{x}(n)|^p - R_p)^2\right] \tag{2.36}$$

where the family parameter p is a positive integer and the Godard parameter R_p is defined as

$$R_p = \frac{\mathrm{E}\left[\left|\left(x_{\mathrm{re}}(n)\right)\right|^{2p}\right]}{\mathrm{E}\left[\left|\left(x_{\mathrm{re}}(n)\right)\right|^p\right]} \,. \tag{2.37}$$

The motivation of Godard's definition of is found in the non-blind cost function

$$J(n) = \mathrm{E}[|\hat{x}(n)|^p - |x(n)|^p]^2 \tag{2.38}$$

for which equalisation is independent of carrier phase. The dispersion function replaces the unobservable input signal with the constant R_p that holds required information about the input distribution. Godard's error signal and nonlinearity are obtained by differentiation of Eq. (2.36)

$$e_G(n) = \left(|\hat{x}(n)|^p - R_p\right)|\hat{x}(n)|^{p-2}\hat{x}(n) \qquad (2.39)$$

and consequently after some mathematics the nonlinearity is found as

$$g_G(\hat{x}(n)) = \frac{\hat{x}(n)}{|\hat{x}(n)|}\left(|\hat{x}(n)| - |\hat{x}(n)|^{2p-1} + R_p|\hat{x}(n)|^{p-1}\right) . \qquad (2.40)$$

The case $p = 1$ is identical to the Sato algorithm, while the case $p = 2$ is known as the constant modulus algorithm (CMA), discovered independently by Treichler and Agee [46]. The widely implemented CMA is regarded as the most successful blind algorithm, due to its low complexity, robustness to carrier phase offset (constant modulus property) and good equalisation performance. The error signal of the CMA is

$$e_{CMA}(n) = (|\hat{x}(n)|^2 - R_2)\hat{x}(n) \qquad (2.41)$$

with Godard parameter

$$R_2 = \frac{E[|x_{re}(n)|^4]}{E[|x_{re}(n)|^2]} . \qquad (2.42)$$

The nonlinearity hence becomes

$$g_{CMA}(\hat{x}(n)) = \hat{x}(n)(1 - |\hat{x}(n)|^2 + R_2) . \qquad (2.43)$$

Godard algorithms for $p > 2$ have higher complexity, without providing any improved performance. As a digression, it should be mentioned that Godard's dispersion function has been generalised to yield a nonlinearity with arbitrarily long memory. Chen et al. [6] proposed an equalisation criterion called Criterion with memory nonlinearity (CRIMNO), that is reported to achieve faster convergence while maintaining low complexity. The CRIMNO cost function is given by

$$J^{(M)}_{CRIMNO}(n) = c_0 E\left[(|\hat{x}(n)|^2 - R_p)^2\right] + \sum_{i=1}^{M} c_i |E[\hat{x}(n)\hat{x}_n^*(n - i)]|^2 \qquad (2.44)$$

where c_i are weights, M denotes the size of the memory and $\hat{x}_n(n - i)$ is the equaliser output at time $(n - i)$ using the equaliser coefficients at time n. Because of the nonzero memory, this is not a Bussgang algorithm.

2.3.5 Benveniste-Goursat Algorithm

Although a number of different Bussgang algorithms have been proposed, most are just variations of the Sato and the Godard algorithm. The algorithm suggested by Benveniste and Goursat (BG) [2] is no exception. The choice of nonlinearity is a heuristic attempt to refine Sato's approach, through combining Sato's error signal with an estimate of the true error $e_t(n)$

$$e_t(n) = \hat{x}(n) - d(n) \qquad (2.45)$$

when $\hat{x}(n)$ is the output of a perfect equaliser with constant delay k and there is a corresponding delay operator in the path of the desired signal. The error signal produced with a nonlinear estimator is sometimes referred to as the pseudo-error signal, as opposed to the true error. The pseudo-error signal has the undesirable property of being noisy around the sought solution for the equaliser. It does not tend to zero even when the Bussgang algorithms converge. This feature is especially distinct for the coarse Sato estimator.

An estimate of the true error can be obtained by a quantiser $Q(\hat{x}(n))$. As defined earlier, this is a decision device that picks the symbol closest to $\hat{x}(n)$ from the symbol constellation of the discrete data source. Hence, the true error estimate is given by

$$e_{DD}(n) = \hat{x}(n) - Q(\hat{x}(n)) \,. \qquad (2.46)$$

This is known as the error signal of the decision-directed (DD) algorithm. Immediately after start-up, $e_{DD}(n)$ will clearly be very erroneous. It is not useful for a purpose of opening up the eye-pattern. Macchi and Eweda [28] have proven that the decision-directed algorithm converges to the optimal tap-setting in the noiseless case, after the eye-pattern has been opened. Only weak capabilities of convergence are exhibited for closed-eye conditions.

Nevertheless, after training the true error estimate will tend to zero, as desired. It is then found that a combination of the Sato error signal $e_S(n)$ and decision-directed error $e_{DD}(n)$ will inherit the benefits of both approaches. The new error signal of Benveniste and Goursat is defined by

$$e_{BG}(n) = k_1 e_{DD}(n) + k_2 |e_{DD}(n)| e_S(n) \qquad (2.47)$$

where k_1 and k_2 are constants that must be adjusted. In the original paper, Benveniste and Goursat use $k_1 = 4$, $k_2 = 1$ for a one carrier system, and $k_1 = 5$, $k_2 = 1$ for a two carrier system. This error signal provides a smooth and automatic switch from start-up mode to decision-directed mode. Conversely, drift in the characteristics of the channel will switch the algorithm back into start-up mode. The BG nonlinearity is

$$g_{BG}(\hat{x}(n)) = \hat{x}(n) - k_1(\hat{x}(n) - Q(\hat{x}(n)))$$
$$- k_2 |\hat{x}(n) - Q(\hat{x}(n))| (\hat{x}(n) - \gamma \operatorname{sign}(\hat{x}(n))) \tag{2.48}$$

with the Sato parameter γ as defined previously.

2.3.6   Stop-and-Go Algorithm

The starting point of Picchi and Pratis 'stop-and-go' algorithm [33] is the decision-directed algorithm, for which the error signal is again given by

$$e_{DD}(n) = \hat{x}(n) - Q(\hat{x}(n)) . \tag{2.49}$$

As outlined under the Benveniste-Goursat algorithm, this error signal can differ significantly from the true error, and will not provide convergence directly from a start-up mode [28]. However, Picchi and Prati demonstrate that a decision-directed algorithm will converge if adaption can be stopped in a fraction of the cases where $e_{DD}(n)$ is unreliable. To obtain this, a binary flag $f(n)$ is introduced. The flag value reports if adaption should be carried out as normal ($f(n) = 1$) or if it should be interrupted ($f(n) = 0$). The formula for updating the filter weights is modified to

$$\mathbf{w}(n) = \mathbf{w}(n-1) - \mu f(n)\mathbf{u}^*(n-k)e(n) . \tag{2.50}$$

We must obtain a criterion for setting the flag values. It is reasonable to assume that if the sign of $e_{DD}(n)$ equals the sign of the true error $e_t(n)$, then the update has the right direction and contributes to convergence. Since the true error is unobservable, another estimate must be used. This estimate is denoted $\hat{e}_t(n)$. $f(n)$ should obey

$$f(n) = \begin{cases} 1, & \text{if } \operatorname{sign}(e_{DD}(n)) = \operatorname{sign}(\hat{e}_t(n)) \\ 0, & \text{if } \operatorname{sign}(e_{DD}(n)) \neq \operatorname{sign}(\hat{e}_t(n)) \end{cases} . \tag{2.51}$$

Thus, $f(n)$ can be expressed as

$$f(n) = \frac{1}{2}\left| \operatorname{sign}(e_{DD}(n)) + \operatorname{sign}(\hat{e}_t(n)) \right| . \tag{2.52}$$

The new error signal for the Stop-and-Go algorithm becomes

$$
\begin{aligned}
e_{SG}(n) &= f(n)\,e_{DD}(n) \\[1ex]
&= \tfrac{1}{2}\left| \operatorname{sign}(e_{DD}(n)) + \operatorname{sign}(\hat{e}_t(n)) \right| e_{DD}(n) \\[1ex]
&= \tfrac{1}{2}\left[\operatorname{sign}(e_{DD}(n)) + \operatorname{sign}(\hat{e}_t(n)) \right] |e_{DD}(n)| \\[1ex]
&= \tfrac{1}{2}\left[e_{DD}(n) + \operatorname{sign}(\hat{e}_t(n))\,|e_{DD}(n)| \right] .
\end{aligned}
\tag{2.53}
$$

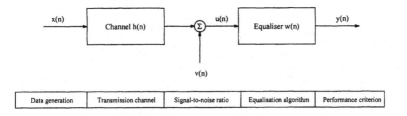

Figure 2.9: Partitioning of Simulation Environment

Proper convergence of the stop-and-go algorithm will depend on a high probability for stopping adaption when signs are unequal. The probability of stopping adaption unnecessarily will primarily have an effect on convergence speed. These numbers will depend on the choice of $\hat{e}_t(n)$, which we have not yet defined. In the original paper, Picchi and Prati argue that a Sato-like error signal should be used. They propose the error signal

$$\hat{e}_t(n) = \hat{x}(n) - \beta_n \text{sign}(\hat{x}(n)) \tag{2.54}$$

where β_n is a coefficient that could possibly be dependent on n. A practical choice is the constant Sato parameter $\beta_n = \gamma$. This gives

$$\hat{e}_t(n) = \hat{x}(n) - \gamma \text{sign}(\hat{x}(n)) . \tag{2.55}$$

The nonlinearity of the stop-and-go algorithm is

$$g_{\text{SG}} = \hat{x}(n) - \frac{1}{2}\Big[\hat{x}(n) - Q\big(\hat{x}(n)\big) \\ + \text{sign}\Big(\hat{x}(n) - \gamma\text{sign}(\hat{x}(n))\Big)\big)\big|\hat{x}(n) - Q\big(\hat{x}(n)\big)\big|\Big] . \tag{2.56}$$

Different formulations of the stop-and-go algorithm are found in the literature. Some combine the decision-directed algorithm and the CMA. Others combine CMA and Sato. This illustrates the variety of possible hybrid solution involving Bussgang algorithms (and other techniques). In any stop-and-go case, the result is an algorithm whose iterative adaption is likely to go in the right direction — when it is allowed to adapt. Thus, a higher step-size parameter can be tolerated during adaption.

2.3.7 Simulation Environment

In this section we want to test and compare the presented Bussgang algorithms. First we need to define different aspects of the simulation environment, as shown in figure 2.9. This environment will also be used to test other classes of algorithms in the sequel.

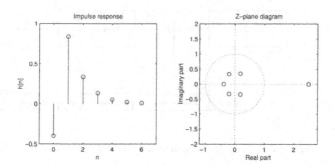

Figure 2.10: Impulse response and z-plane zero diagram of easy channel

Transmitted data is modulated with two different modulation schemes, QAM-16 and the V.29 standard. Sequences of i.i.d. symbol are produced with random generator functions modelling these sources.

Two non-minimum phase channels are selected to test the blind equalisation algorithms. One has zeros far from the unit circle of the z-plane, and should be relatively easy to equalise. The other channel exhibits more awkward zero locations, producing a deep notch in its magnitude response. These will be referred to as the "easy" and the "difficult" channel.

Both easy [38] and difficult [39] channel have been used in simulations described in the papers by Shalvi and Weinstein. Impulse responses and z-plane diagrams of the chosen channels are shown in figure 2.10 and 2.12. Frequency responses are shown in figure 2.11 and 2.13. White zero-mean Gaussian noise is generated and added to the channel output. The noise variance is determined to produce a signal-to-noise (SNR) ratio of 30 dB. SNR is defined as

$$ \text{SNR} = 10 \log \frac{E[|x(n)|^2]}{E[|\eta(n)|^2]} = 10 \log \frac{\sigma_x^2}{\sigma_\eta^2} \, . \tag{2.57} $$

For the QAM-16 source the noise variance evaluates to $\sigma_\nu^2 = 1.00 \times 10^{-2}$. For the V.29 source the result is $\sigma_\nu^2 = 1.35 \times 10^{-2}$.

2.3.8 Simulations with Bussgang Algorithms

Equalisation performance is measured with the intersymbol interference defined previously. ISI is measured and averaged over 5 realisations with sequences of 5,000 (easy channel) and 10,000 symbols (difficult channel). For the easy channel, equalisers have 6 filter taps and are initialised by setting

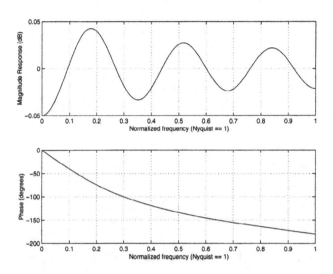

Figure 2.11: Frequency response of easy channel

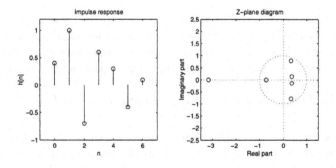

Figure 2.12: Impulse response and z-plane zero diagram of difficult channel

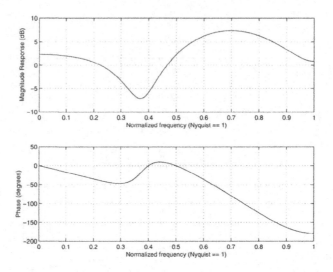

Figure 2.13: Frequency response of difficult channel

the third tap equal to one and all others to zero. For the difficult channel, 24 taps were employed to obtain decent results. Initially, filter tap number twelve was set to one.

For a fair comparison of the algorithms, the step-size parameter μ must be optimised in each single case. Step-size parameters for individual realisations are shown in table 2.1. Other constants are the Sato parameter, the Godard parameter for $p = 2$ and the Benveniste-Goursat parameters. For QAM-16 data we have $\gamma = 2.5$ and $R_2 = 8.1$. For V.29 we have $\gamma = 3.375$ and $R_2 = 19.15$. Independent of modulation scheme, $k_1 = 4$ and $k_2 = 1$.

Simulations with easy and difficult channel are shown respectively in figure 2.14 and 2.15. It is observed that the performance of the algorithms are significantly worse for V.29 data than for QAM-16 data. When V.29 data is used with the difficult channel, ill-convergence is dominant. Only the Godard algorithm managed to equalise the channel, but only for two of the five realisations. In these cases, a final ISI of -14 dB was reached, while the plot illustrates the inadequate average performance.

Table 2.1: Step-size parameters μ in simulation.

	Easy channel		Difficult channel	
	QAM-16	V.29	QAM-16	V.29
CMA	$1 \cdot 10^{-4}$	$2 \cdot 10^{-5}$	$2 \cdot 10^{-5}$	$1 \cdot 10^{-5}$
Sato	$5 \cdot 10^{-4}$	$8 \cdot 10^{-4}$	$2 \cdot 10^{-4}$	$3 \cdot 10^{-4}$
BG	$4 \cdot 10^{-4}$	$4 \cdot 10^{-4}$	$1 \cdot 10^{-4}$	$1 \cdot 10^{-4}$
SG	$1 \cdot 10^{-3}$	$2 \cdot 10^{-3}$	$8 \cdot 10^{-4}$	$1 \cdot 10^{-3}$

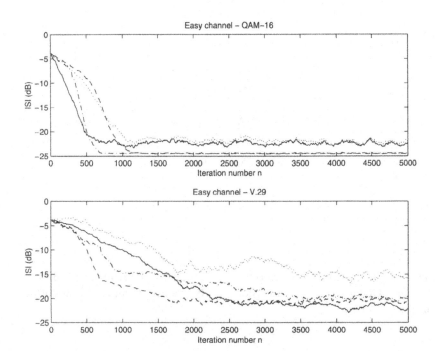

Figure 2.14: Equalisation performance of Bussgang algorithms with easy channel: Godard (solid), Sato (dotted), Benveniste-Goursat (dash-dotted) and Stop-and-Go (dashed)

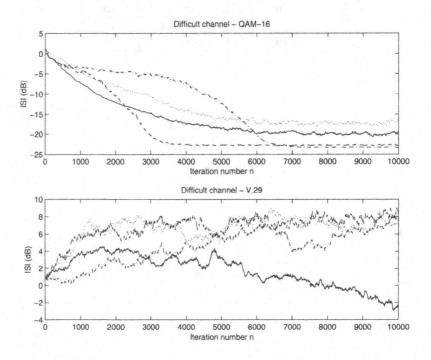

Figure 2.15: Equalisation performance of Bussgang algorithms with difficult channel: Godard (solid), Sato (dotted), Benveniste-Goursat (dash-dotted) and Stop-and-Go

2.4 Algorithms Based on Explicit HOS

The gradient descent algorithms in the previous section do not involve any explicit computation of higher-order statistics (HOS). Nevertheless, a study of the algorithms in the s-domain show that they approach the condition of the earlier stated equalisation criterion, when they converge. This can only be accomplished by somehow resorting to HOS. Therefore, the gradient search algorithms are said to be implicitly based on HOS.

The class of algorithms explicitly based on HOS use statistics like higher-order cumulants, cepstra and their Fourier transforms (polyspectra) to reveal nonminimum phase characteristics of the unknown channel. Such methods relate HOS of the received signal directly to the unknown channel and sought equaliser. They have evolved after the presented Bussgang algorithms, and outperform the previous on convergence rate. The price paid is that explicit HOS-based methods demand more computational power to compute empirical statistics.

Three different algorithms are presented in this section. The tricepstrum equalisation algorithm (TEA) [32] [20] is an early approach suggested by Hatzinakos and Nikias [20], using the cepstrum of the fourth-order cumulant. While being one of the blockbusters demonstrating the possible achievements of explicit HOS-based methods, this algorithm is hardly of practical importance anymore.

The Benveniste-Goursat-Ruget theorem [3] proves that perfect equalisation is obtained when the pdf of the equaliser output is identical to the known pdf of the channel input. A relaxed condition was given by Shalvi and Weinstein, who showed that it is sufficient to equalise only a few statistics, and coined the important maximum kurtosis criterion [38]. Their work led to the super-exponential algorithm (SEA) [39] as well as Jellonek and Kammeyer's eigenvector approach (EVA) [24,25]. The latter algorithms are more appealing than the TEA in terms of both simplicity and performance.

It should also be noted that the TEA is essentially a blind identification procedure. It goes through intermediate step of channel identification, before determining the equaliser. Conversely, the SEA and EVA are pure equalisation schemes that estimate the inverse channel directly from the empirical statistics. The nonlinearity of the explicit HOS-based algorithms is a HOS-estimator. In the TEA, the nonlinearity takes only the received signal as output. The SEA and EVA nonlinearities also use the equaliser output. This is illustrated in figure 2.16 which models a basic blind equalisation configuration for explicit HOS-based algorithms.

Other explicit HOS-based algorithms include two parametric methods using second- and fourth-order moments, proposed by Porat and Friedlan-

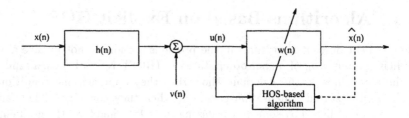

Figure 2.16: Block diagram of general HOS-based equaliser

der [34]. The first is based on the least-squares solution of a linear set of equations, while the second uses a nonlinear procedure to minimise a certain HOS-based cost function.

2.4.1 Tricepstrum Equalisation Algorithm

Definition of Polycepstra

Preliminary to the derivation of the TEA, we have to introduce the higher-order complex cepstrum. Let $C_{nx}(\omega_1, \ldots, \omega_{n-1})$ be the nth order cumulant spectrum of the random sequence $x(n)$. Then, the nth order cepstrum is defined as

$$\kappa_{nx}(k_1, \ldots, k_{n-1}) = \mathfrak{F}^{-1} \{\ln C_{nx}(\omega_1, \ldots, \omega_{n-1})\} . \tag{2.58}$$

where $\mathfrak{F}^{-1}$ is the inverse $(n-1)$-dimensional Fourier transform. For our application, we are interested in the tricepstrum (fourth-order cepstrum) of the channel output, which is given by

$$\kappa_{4u}(k_1, k_2, k_3) = \mathfrak{F}^{-1} \{\ln C_{4u}(\omega_1, \omega_2, \omega_3)\} . \tag{2.59}$$

With a linear time-invariant and causal channel $h(n)$, the following equivalent relationships are valid for the cumulant and cumulant spectrum

$$c_{4u}(k_1, k_2, k_3) = c_{4x}(k_1, k_2, k_3) \sum_{n=0}^{+\infty} h_n h_{n+k_1} h_{n+k_2} h_{n+k_3} , \tag{2.60}$$

$$C_{4u}(\omega_1, \omega_2, \omega_3) =$$
$$= C_{4x}(\omega_1, \omega_2, \omega_3) H(\omega_1) H(\omega_2) H(\omega_3) H(-\omega_1 - \omega_2 - \omega_3) . \tag{2.61}$$

Hence, evaluation of the trispectrum requires calculation of

$$\ln C_{4u}(\omega_1, \omega_2, \omega_3) = \ln C_{4x}(\omega_1, \omega_2, \omega_3) + \ln H(\omega_1) + \ln H(\omega_2)$$
$$+ \ln H(\omega_3) + \ln H(-\omega_1 - \omega_2 - \omega_3) . \tag{2.62}$$

Decomposition of Non-minimum Phase Channel

$H(\omega)$ is a non-minimum phase system that can be decomposed into one part with zeros within and one part with zeros outside the unit circle of the z-plane $(z = e^{j\omega})$

$$H(\omega) = k\, I(\omega)\, O(-\omega)$$
$$= k\, \prod_{l=1}^{L_1}(1 - a_l e^{-j\omega}) \prod_{l=1}^{L_2}(1 - b_l e^{j\omega}) \,. \tag{2.63}$$

This leads to

$$\ln H(\omega) = \ln k + \ln I(\omega) + \ln O(-\omega)$$
$$= \ln k + \sum_{l=1}^{L_1}\ln\left(1 - a_l e^{-j\omega}\right) + \sum_{l=1}^{L_2}\ln\left(1 - b_l e^{j\omega}\right) \tag{2.64}$$

and

$$\ln H(-\omega_1 - \omega_2 - \omega_3) = \ln k + \sum_{l=1}^{L_1}\ln\left(1 - a_l e^{j(\omega_1+\omega_2+\omega_3)}\right)$$
$$+ \sum_{l=1}^{L_2}\ln\left(1 - b_l e^{-j(\omega_1+\omega_2+\omega_3)}\right) \,. \tag{2.65}$$

With these expressions substituted into Eq. (2.59), an inverse Fourier transform yields the trispectrum

$$\kappa_{4u}(k_1, k_2, k_3) = \begin{cases} \ln k + 3\ln c_{4x}(k_1, k_2, k_3)\,, & k_1 = k_2 = k_3 = 0\,, \\ -\frac{1}{k_1} A^{(k_1)}\,, & k_1 > 0,\ k_2 = k_3 = 0\,, \\ -\frac{1}{k_2} A^{(k_2)}\,, & k_2 > 0,\ k_1 = k_3 = 0\,, \\ -\frac{1}{k_3} A^{(k_3)}\,, & k_3 > 0,\ k_1 = k_2 = 0\,, \\ \frac{1}{k_1} B^{(k_1)}\,, & k_1 < 0,\ k_2 = k_3 = 0\,, \\ \frac{1}{k_2} B^{(k_2)}\,, & k_2 < 0,\ k_1 = k_3 = 0\,, \\ \frac{1}{k_3} B^{(k_3)}\,, & k_3 < 0,\ k_1 = k_2 = 0\,, \\ -\frac{1}{k_2} B^{(k_2)}\,, & k_1 = k_2 = k_3 > 0\,, \\ \frac{1}{k_2} A^{(k_2)}\,, & k_1 = k_2 = k_3 < 0\,, \\ 0\,, & \text{otherwise} \end{cases} \tag{2.66}$$

where

$$A^{(m)} = \sum_{l=1}^{L_1} a_l^m \qquad \text{and} \qquad B^{(m)} = \sum_{l=1}^{L_2} b_l^m \,.$$

Channel Estimation

The coefficients $A^{(m)}$ and $B^{(m)}$ contain respective information about the minimum-phase and the maximum-phase part of the channel. They are defined for a wide (possibly infinite) range of m, but decay exponentially and therefore only a limited number of coefficients needs to be calculated.

Further, the cepstrum parameters $A^{(m)}$ and $B^{(m)}$ are related to the minimum phase time sequence $i(n)$ and the maximum-phase time sequence $o(n)$ as follows

$$i(n) = \begin{cases} -\frac{1}{n}\sum_{m=1}^{n} A^{(m)}i(n-m) , & n = 1, 2, \ldots, L_1 , \\ 1 , & n = 0 , \\ 0 , & n < 0 . \end{cases} \tag{2.67}$$

$$o(n) = \begin{cases} \frac{1}{n}\sum_{m=n}^{-1} B^{(-m)}o(n-m) , & n = -1, -2, \ldots, -L_2 , \\ 1 , & n = 0 , \\ 0 , & n > 0 . \end{cases} \tag{2.68}$$

From Eq. (2.63) we know that the unknown channel is characterised by a linear convolution of the minimum-phase and maximum-phase time sequences, scaled by a constant gain. The estimate of the channel impulse response is therefore

$$\hat{h}(n) = k\,\hat{i}(n) * \hat{o}(n) \tag{2.69}$$

where the gain k is yet to be determined. The problem is now to obtain estimates of the tricepstrum parameters. This can be done by relating the tricepstrum to the fourth-order cumulant through

$$\sum_{r=-\infty}^{+\infty} \sum_{s=-\infty}^{+\infty} \sum_{t=-\infty}^{+\infty} r\,\kappa_{4u}(r, s, t)c_{4u}(k_1 - r, k_2 - s, k_3 - t) =$$
$$= -k_1\,c_{4u}(k_1, k_2, k_3) . \tag{2.70}$$

When substituting the tricepstrum expression of Eq. (2.66) into the previous expression, the result is the tricepstral equation

$$\sum_{m=1}^{p} \Big(A^{(m)}\big(c_{4u}(k_1 - m, k_2, k_3) - c_{4u}(k_1 + m, k_2 + m, k_3 + m)\big)\Big)$$
$$+ \sum_{m=1}^{q} \Big(B^{(m)}\big(c_{4u}(k_1 - m, k_2 - m, k_3 - m) - c_{4u}(k_1 + m, k_2, k_3)\big)\Big) =$$
$$= -k_1\,c_{4u}(k_1, k_2, k_3) . \tag{2.71}$$

The equation describes an overdetermined system of a theoretically infinite number of equations ($q, p \to \infty$). But due to the exponential decay of the tricepstrum parameters, the range of p and q can be truncated.

Parameters p and q are chosen so that only significant $A^{(m)}$ and $B^{(m)}$ are used. Then cumulants are estimated for a finite range of arguments k_1, k_2, k_3. The equation system to be solved can be formulated as

$$\mathbf{Ca} = \mathbf{p} \tag{2.72}$$

where the $N_c \times (p+q)$ matrix $\mathbf{C}$ contains the left-hand ($c_{4u}(\cdot) - c_{4u}(\cdot)$)-terms of Eq. (2.71), the $(p+q) \times 1$ vector $\mathbf{a}$ stores the tricepstrum-parameters $A^{(m)}$ and $B^{(m)}$, while the $N_c \times 1$ vector $\mathbf{p}$ holds the right-hand $-k_1\, c_{4u}(k_1, k_2, k_3)$-terms.

Estimates of $\mathbf{C}$ and $\mathbf{p}$ are used to solve for $\hat{\mathbf{a}}$. The entries of $\hat{\mathbf{C}}$ and $\hat{\mathbf{p}}$ are formed of cumulant estimates that are calculated from the channel output in each iteration. A normal solution for $\hat{\mathbf{a}}$ is obtained from

$$\hat{\mathbf{a}} = \hat{\mathbf{C}}^{+}\hat{\mathbf{p}} = \left(\hat{\mathbf{C}}^{T}\hat{\mathbf{C}}\right)^{-1}\hat{\mathbf{C}}^{T}\hat{\mathbf{p}} \tag{2.73}$$

where $\hat{\mathbf{C}}^{+}$ is the pseudo-inverse of matrix $\hat{\mathbf{C}}$. The channel estimate $\hat{h}[n]$ is calculated from Eq. (2.69), after elements of $\hat{\mathbf{a}}$ are used to determine $\hat{\imath}\,[n]$ and $\hat{o}[n]$ through Eq. (2.67) and (2.68). When a channel estimate is available, the equaliser is estimated directly from the equalisation criterion

$$\hat{h}(n) * \hat{w}(n) = \delta(n - k)e^{j\theta} \; . \tag{2.74}$$

In the frequency domain this looks like

$$\hat{W}(\omega) = \frac{1}{\hat{H}(\omega)}e^{j(\theta - k\omega)} \tag{2.75}$$

where $\hat{W}(\omega)$ is the Fourier transform of the equaliser estimate. Thus, $\hat{w}(n)$ is obtained from the inverse transform.

2.4.2 Maximum Kurtosis Criterion

Remember that in the gradient descent algorithms the output signal is obtained from a nonlinear estimator

$$\hat{d}(n) = g\big(\hat{x}(n)\big) \approx x(n) \qquad \text{with} \qquad \hat{x}(n) = x(n) + \eta(n) \; .$$

The superimposed convolution noise $\eta(n)$ at the final stage is independent of the signal, Gaussian and mean free. Based on the mutual independence and Gaussianity, the application of HOS (higher than second-order) should be obvious due to the insensitivity to Gaussian noise.

Respectively a rather new approach in blind equalisation is based on the so-called maximum kurtosis criterion. The fundamental idea behind is to exploit the Schwartz inequality

$$\sum_k |s(k)|^4 \leq \left(\sum_k |s(k)|^2 \right)^2$$

which holds the equality for only one nonzero element of **s**. This is also the condition for perfect equalisation. The application of this theorem to cumulants is straightforward, recalling that for an i.i.d. input sequence we have

$$c_{2\hat{x}}(0) = c_{2x}(0) \sum_k |s(k)|^2 \ ,$$

$$c_{4\hat{x}}(0,0,0) = c_{4x}(0,0,0) \sum_k |s(k)|^4 \ . \tag{2.76}$$

Both latter equations follow from derivations using elementary cumulant properties. Remark also that kurtosis and variance of input and output process must have the same sign. The cumulants are inserted in the inequality to obtain

$$\frac{c_{4\hat{x}}(0,0,0)}{c_{4x}(0,0,0)} \leq \left(\frac{c_{2\hat{x}}(0)}{c_{2x}(0)} \right)^2 \ . \tag{2.77}$$

With the additional conditions

$$c_{2\hat{x}}^2(0) = \alpha^4 c_{2x}^2(0) \quad \text{and} \quad c_{4\hat{x}}(0,0,0) = \alpha^4 c_{4x}(0,0,0)$$

where α^2 is an arbitrary power gain through the system, the maximisation of the absolute value of the output kurtosis $|c_{4\hat{x}}(0,0,0)|$ must finally lead to the desired aim as long as the right-hand side of the inequality stays constant.

2.4.3   Super-Exponential Algorithm

We now turn to the application of the maximum kurtosis criterion. The simultaneous 2nd and 4th order cumulant of the system output $\hat{x}(n)$ are defined as

$$c_{2\hat{x}}(0) = \mathrm{E}\big[|\hat{x}(n)|^2\big] \ ,$$

$$c_{4\hat{x}}(0,0,0) = \mathrm{E}\big[|\hat{x}(n)|^4\big] - 2\Big(\mathrm{E}\big[|\hat{x}(n)|^2\big] \Big)^2 - \Big| \mathrm{E}\big[\hat{x}^2(n)\big] \Big|^2 \ . \tag{2.78}$$

Extremisation of the kurtosis function must be performed under the purely real constraint

$$\mathcal{C} = \alpha^4 c_{2x}^2(0) - c_{2\hat{x}}^2(0)$$
$$= c_{2x}^2(0)\left(\alpha^4 - \left[\sum_k |s(k)|^2\right]^2\right). \tag{2.79}$$

Constrained extremisation is handled by the procedure known as the method of Lagrange multipliers. The chosen constraint normalises the power gain trough the system to the arbitrary constant α^2. The appropriate Lagrangian that we aim to maximise is obtained as

$$\mathcal{L} = \left|c_{4\hat{x}}(0,0,0)\right| + \lambda\mathcal{C}$$
$$= \mathrm{E}\left[|\hat{x}(n)|^4\right] - 2\left(\mathrm{E}\left[|\hat{x}(n)|^2\right]\right)^2 - \left|\mathrm{E}\left[\hat{x}^2(n)\right]\right|^2 \tag{2.80}$$
$$+ \lambda\left(\alpha^4 c_{2x}^2(0) - \left[\mathrm{E}\left[|\hat{x}(n)|^2\right]\right]^2\right).$$

Here and for the following considerations the absolute value operator as it appears in the first line is disregarded firstly and is included at a later instant.

First-order partial differentiation with respect to **w** gives

$$\nabla_{\mathbf{w}}\mathcal{L} = 2\mathrm{E}\left[|\hat{x}(n)|^2\hat{x}^*(n)\mathbf{u}^T(n)\right] - 4\mathrm{E}\left[|\hat{x}(n)|^2\right]\mathrm{E}\left[\hat{x}^*(n)\mathbf{u}^T(n)\right]$$
$$- 2\mathrm{E}\left[(\hat{x}^*(n))^2\right]\mathrm{E}\left[\hat{x}(n)\mathbf{u}^T(n)\right] - 2\lambda c_{2\hat{x}}(0)\mathbf{w}^{*T}\mathrm{E}\left[\mathbf{u}^*(n)\mathbf{u}^T(n)\right]$$
$$= 2\mathbf{c}_{3\hat{x}\mathbf{u}}^T - 2\lambda c_{\hat{x}}(0)\mathbf{w}^{*T}\mathbf{C}_{2\mathbf{u}} . \tag{2.81}$$

where we define the joint cumulant vector

$$\mathbf{c}_{3\hat{x}\mathbf{u}} = \begin{bmatrix} c_{3\hat{x}u}(0,0,0) & \cdots & c_{3\hat{x}u}(0,0,-L) \end{bmatrix}^T \tag{2.82}$$

and the correlation matrix

$$\mathbf{C}_{2\mathbf{u}} = \begin{bmatrix} c_{2u}(0) & \cdots & c_{2u}(-L) \\ \vdots & & \vdots \\ c_{2u}(L) & \cdots & c_{2u}(0) \end{bmatrix} . \tag{2.83}$$

Here is taken advantage from the fact that the signal $\hat{x}(n) = \mathbf{u}^T\mathbf{w}$ contains the linear combination of the equaliser filter as a deterministic part and the random equaliser input. The deterministic filter vector can be extracted from the expectation operation. Partial differentiation with respect to λ yields the constraint itself

$$\frac{\partial}{\partial\lambda}\mathcal{L} = \alpha^4 c_{2x}^2(0) - c_{2\hat{x}}^2(0) .$$

From the definition it is assumed that an unique extremum, the global maximum, exists. The mathematical proof here is left out. The minimum of the fundamental cost function is obtained by setting the gradient to zero. The stationary point defined by

$$\nabla_{\mathbf{w}}\mathcal{L} = \mathbf{0}^T \tag{2.84}$$

is called the SEA equilibrium, from the super-exponential algorithm of Shalvi and Weinstein [39]. The SEA solution is found from

$$\mathbf{c}_{3\hat{x}\mathbf{u}}^T - \lambda c_{2\hat{x}}(0)\mathbf{w}^{*T}\mathbf{C}_{2\mathbf{u}} = \mathbf{0}^T . \tag{2.85}$$

The resulting expression can be exploited to solve for the equaliser filter vector

$$\mathbf{w} = \frac{1}{\lambda c_{2\hat{x}}(0)}\left(\mathbf{c}_{3\hat{x}\mathbf{u}}^T\mathbf{C}_{2\mathbf{u}}^{-1}\right)^{*T} . \tag{2.86}$$

An appropriate cumulant-based method can now be developed [39] ([38]). The effective value of $\lambda c_{2\hat{x}}(0)$ affects merely the gain of the overall system and does not really matter as long as it is not equal to zero. To show that this is the case, Eq. (2.85) is reinserted into the unconstrained ($\lambda\mathcal{C} = 0$) kurtosis function in Eq. (2.80). The earlier dropped absolute value operator is here applied again

$$\begin{aligned}
\left|c_{4\hat{x}}(0,0,0)\right| &= \left|\mathbf{c}_{3\hat{x}\mathbf{u}}^T\mathbf{w}\right| \\
&= \left|\lambda c_{2\hat{x}}(0)\mathbf{w}^{*T}\mathbf{C}_{2\mathbf{u}}\mathbf{w}\right| \\
|\lambda| &= \frac{\left|c_{4\hat{x}}(0,0,0)\right|}{c_{2\hat{x}}^2(0)} .
\end{aligned} \tag{2.87}$$

2.4.4 Eigenvector Approach

The solution of the eigenvector approach (EVA) [24, 25] can be derived in a straightforward manner from the SEA equilibrium. The cumulant vector is changed into a matrix by extracting a reference equaliser filter vector $\mathbf{w}$ from the cumulant expression as it was done before

$$\begin{aligned}
\mathbf{0} &= 2\mathbf{w}^{*T}\mathrm{E}\big[|\hat{x}(n)|^2\mathbf{u}^*(n)\mathbf{u}^T(n)\big] - 2\mathbf{w}^{*T}\mathrm{E}\big[|\hat{x}(n)|^2\big]\,\mathrm{E}\big[\mathbf{u}^*(n)\mathbf{u}^T(n)\big] \\
&\quad - 2\mathbf{w}^{*T}\mathrm{E}[\hat{x}(n)\mathbf{u}^*(n)]\,\mathrm{E}[\hat{x}^*(n)\mathbf{u}^T(n)] \\
&\quad - 2\mathbf{w}^{*T}\mathrm{E}[\hat{x}^*(n)\mathbf{u}^*(n)]\,\mathrm{E}[\hat{x}(n)\mathbf{u}^T(n)] - 2\lambda c_{2\hat{x}}(0)\mathbf{w}^{*T}\mathbf{C}_{2\mathbf{u}} \\
&= 2\mathbf{w}^{*T}\mathbf{C}_{2\hat{x}2\mathbf{u}} - 2\lambda c_{2\hat{x}}(0)\mathbf{w}^{*T}\mathbf{C}_{2\mathbf{u}} .
\end{aligned}$$

The simplified cumulant matrix notation used above is defined as

$$\mathbf{C}_{2\hat{x}2u} = \begin{bmatrix} c_{2\hat{x}2u}(0,0,0) & \cdots & c_{2\hat{x}2u}(0,0,-L) \\ \vdots & & \vdots \\ c_{2\hat{x}2u}(0,-L,0) & \cdots & c_{2\hat{x}2u}(0,-L,-L) \end{bmatrix} . \tag{2.88}$$

The obtained equation is recognised as a generalised eigenvector problem, which defines the EVA solution

$$\left(\lambda c_{2\hat{x}}(0)\right)\mathbf{C}_{2u}\mathbf{w} = \mathbf{C}_{2\hat{x}2u}\mathbf{w} \qquad \text{with} \qquad \mathbf{C}_{2u} = \mathbf{C}_{2u}^{*T} , \ \mathbf{C}_{2\hat{x}2u} = \mathbf{C}_{2\hat{x}2u}^{*T} . \tag{2.89}$$

As indicated in Eq. (2.87), the eigenvector $\left(\lambda c_{2\hat{x}}(0)\right)$ associated with the maximum eigenvalue by the same sign as the cumulant must be chosen to enforce the desired equality in Eq. (2.77).

The entire procedure has to be repeated several times. However, after only a few iterations and respective updates of the equaliser output, the desired solution will be approached. As usual in blind equalisation applications the arbitrary gain α cannot be determined. This can be bypassed by an additional automatic gain control. In anticipation of the simulation results it can be said that, comparing SEA and EVA, the EVA initially converges slightly faster, but it also increases the computational load because of the required eigenvalue decomposition.

Exemplary for the EVA, the complete algorithms is summarised in the following steps:

1. Set up the initial equaliser with only one non-zero element. The position of this element determines mostly the resulting delay trough the overall system and should be chosen in a central positions if no additional knowledge is available

$$\mathbf{w}_{\text{initial}} = \begin{pmatrix} 0 & \cdots & 0 & 1 & 0 & \cdots & 0 \end{pmatrix}^{T} .$$

2. Calculate the equaliser output $\hat{x}(n)$ for $n = L+1, \ldots, N$

$$\hat{x}(n) = \mathbf{w}^{T}\mathbf{u}(n) .$$

3. Estimate the correlation matrix

$$\mathbf{C}_{2u} = \frac{1}{N-L} \sum_{n=L+1}^{N} \mathbf{u}^{*}(n)\mathbf{u}^{T}(n)$$

and cumulant matrix

$$
\mathbf{C}_{2\hat{x}2u} = \frac{1}{N-L} \sum_{n=L+1}^{N} \hat{x}(n)\hat{x}^*(n)\mathbf{u}(n)\mathbf{u}^T(n)
$$

$$
- \frac{1}{(N-L)^2} \left(\sum_{n=L+1}^{N} \hat{x}(n)\hat{x}^*(n) \sum_{n=L+1}^{N} \mathbf{u}^*(n)\mathbf{u}^T(n) + \right.
$$

$$
\sum_{n=L+1}^{N} \hat{x}(n)\mathbf{u}^*(n) \sum_{n=L+1}^{N} \hat{x}^*(n)\mathbf{u}^T(n) +
$$

$$
\left. \sum_{n=L+1}^{N} \hat{x}^*\mathbf{u}^*(n) \sum_{n=L+1}^{N} \hat{x}(n)\mathbf{u}^T(n) \right)
$$

as averages over a stationary time interval. The cumulant average can be simplified in various manners, which is not done here to stay more general.

4. Apply the generalised eigenvalue decomposition to solve for the maximum eigenvalue $|\lambda c_{2\hat{x}}(0)|_{\max}$

$$
\left(\lambda c_{2\hat{x}}(0)\right) \mathbf{C}_{2u}\mathbf{w} = \mathbf{C}_{2\hat{x}2u}\mathbf{w}
$$

or equivalently apply the eigenvalue decomposition

$$
\left(\lambda c_{2\hat{x}}(0)\right)\mathbf{w} = \left(\mathbf{C}_{2u}^{-1}\mathbf{C}_{2\hat{x}2u}\right)\mathbf{w} \ .
$$

The respective equaliser filter is the with $|\lambda c_{2\hat{x}}(0)|_{\max}$ associated eigenvector $\mathbf{w}$.

5. Go back to step 2 and repeat until a sufficient equalisation result is obtained or the equaliser filter does not change anymore. This should quickly occur after only a few iterations (see simulations in subsection 2.4.6).

2.4.5 Adaption by First-Order Stochastic Approximation.

First-order stochastic approximations of expectation operations are often used in the adaptive filtering context. It allows to operate with the currently incoming input samples without a large overhead and the connected known problems such as an additional memory to store the required samples and a large initial response time. The task of first-order stochastic approximation is performed by replacing the true gradient of the cost function by its sample estimate.

The exploitation of the gradient expression in Eq. (2.81) for a first-order stochastic approximation approach is given by

$$(\nabla_{\mathbf{w}}\mathcal{L})^{*T} = 2\mathbf{u}^*(n)\Big(|\hat{x}(n)|^2\hat{x}(n) - 2\underbrace{\mathrm{E}\big[|\hat{x}(n)|^2\big]}_{\alpha_1}\hat{x}(n)$$

$$- \underbrace{\mathrm{E}\big[\hat{x}^2(n)\big]}_{\alpha_2}\hat{x}^*(n) - \lambda c_{2\hat{x}}(0)\hat{x}(n)\Big) .$$

All expectation operations are removed and only the values α_1 and α_2 are treated as initial known constant values. For the important class of complex, symmetrically distributed signals

- the moment $\mathrm{E}[x^2(n)]$ equals zero and therefore $\alpha_2 = 0$.

From this follows

$$(\nabla_{\mathbf{w}}\mathcal{L})^* = 2\mathbf{u}^*(n)\hat{x}(n)\Big(|\hat{x}(n)|^2 - 2\alpha_1 - \lambda c_{2\hat{x}}(0)\Big)$$

- and with the definition of the constant R_2

$$2\alpha_1 + \lambda c_{2\hat{x}}(0) = \frac{\mathrm{E}[|x(n)|^4]}{\mathrm{E}[|x(n)|^2]} = R_2$$

the gradient term of the CMA algorithm is recognised (compare Eq. (2.41))

$$(\nabla_{\mathbf{w}}\mathcal{L})^* = \mathbf{u}^*(n)\underbrace{\hat{x}(n)\Big(|\hat{x}(n)|^2 - R_2\Big)}_{e_{\mathrm{CMA}}(n)} .$$

The gradient expression can be used in the gradient search (steepest descent method)

$$\mathbf{w}(n) = \mathbf{w}(n-1) - \mu(\nabla_{\mathbf{w}}\mathcal{L})^* .$$

The Lagrange multiplier λ can be obtained as

$$\lambda = \frac{\mathrm{E}[\{|x(n)|^4\}] - 2\Big(\mathrm{E}[\{|x(n)|^2\}]\Big)^2}{\Big(\mathrm{E}[\{|\hat{x}(n)|^2\}]\Big)^2} = \frac{c_{4x}(0,0,0)}{c_{2\hat{x}}^2(0)}$$

which should not be surprising after Eq. (2.87). As known before, equalisation of Gaussian distributed input signals is not possible in this way.

To sum it up it can be said that the earlier introduced Godard or constant modulus algorithm (CMA) can be seen as first-order stochastic approximation of the super-exponential approach. This fact can support analysis and classification of the CMA for different types of signals.

2.4.6   Simulations with Explicit HOS-Based Algorithms

To show the performance of the previously introduced algorithms some simulations are done. The simulation result are averages of five independent runs. To allow the competition of the multiple channel and the single channel the MSE is chosen as a evaluation criterion instead of the earlier used ISI. For the underlying noise-free case ISI and MSE are related as follows

$$
\begin{aligned}
\mathrm{E}\big[\{|\epsilon(n)|^2\}\big] &= \mathrm{E}\Big[\big\{|\hat{x}(n) - \alpha x(n - n_0)|^2\big\}\Big] \\
&= \mathrm{E}\Bigg[\Big\{\Big|\sum_{d=0}^{K+L}\big(s(d) - \alpha\delta(d - n_0)\big)x(n - d)\Big|^2\Big\}\Bigg] \\
&= \sum_{d_1}\sum_{d_2}\big(s(d_1) - \alpha\delta(d_1 - n_0)\big)\big(s(d_2) - \alpha\delta(d_2 - n_0)\big)^* \\
&\qquad \mathrm{E}[\{x(n - d_1)x^*(n - d_2)\}] \\
&= \underbrace{\mathrm{E}\big[\{|x(n)|^2\}\big]}_{c_{2x}(0)}\Big(\sum_d|s(d)|^2 - \alpha^* s_{\mathrm{max}} - \alpha s_{\mathrm{max}}^* + |\alpha|^2\Big).
\end{aligned}
$$

To obtain the last line the i.i.d. assumption is applied. The factor α is defined in a way that $\alpha = s_{\mathrm{max}}$. Note, that in this case

$$
\mathrm{E}\big[|e(n)|^2\big] = c_{2x}(0)|s_{\mathrm{max}}|^2\frac{\sum_d|s(d)|^2 - |s_{\mathrm{max}}|^2}{|s_{\mathrm{max}}|^2}
$$

is obtained where the quotient defines the ISI.

For the following simulations, the normalised MSE

$$
(\mathrm{N})\mathrm{MSE} = \frac{\mathrm{E}[|e(n)|^2]}{c_{2x}(0)|s_{\mathrm{max}}|^2}
$$

is used, which directly equals the ISI and removes the dependence of the MSE on the input variance and on a possible gain of the overall system.

$N = 300$ input samples noise free

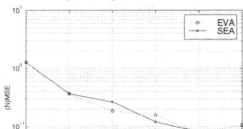

The simulation shows the wide similarity in the performance as expected.

Conditions: QAM4 input signal with $N = 300$ samples, noise free and additive noise SNR=30dB, equaliser filter with $L + 1 = 7$ coefficients for the "difficult" channel type.

Figure 2.17: MSE behaviour of EVA and SEA within the iterations

The behaviour of the mean square error for EVA and SEA is shown in figures 2.17, 2.18 and 2.19 with respect to the number of iterations and the effects of additive noise. The graphs show, that the final MSE is obtained after a few iterations. As expected the algorithms do not differ significantly, because of their common mathematical foundation.

2.5 Equalisation with Multiple Channels

2.5.1 Fractionally Spaced Equalisation

Early adaptive equalisers were all symbol rate spaced (or T-spaced). This means that the transmitted signal is a stream of discrete symbol spaced in time by the symbol period T, and that the received signal is sampled at the corresponding reciprocal $1/T$, known as the baud rate. The alternative that is discussed in this subsection is a concept called fractionally spaced equalisation, which implies that the equaliser input is sampled faster the the symbol rate.

Signal acquisition through oversampling by an integer multiple of the original baud rate is called fractional sampling. It can be achieved by taking more than one sample per symbol period. Samples taken at each slot of the original symbol period are then processed separately by individual equalisers,

$N = 300$ input samples, additive noise with SNR=30dB

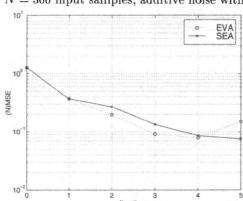

The application of an additive noise does not influence the performance of the algorithms so far as the noise floor is below the MSE of the final convergence state.

Conditions: QAM4 input signal with $N = 300$ samples, additive noise SNR=30dB, equaliser filter with $L + 1 = 7$ coefficients for the "difficult" channel.

Figure 2.18: MSE behaviour of EVA and SEA within the iterations

$N = 800$ input samples, noise free

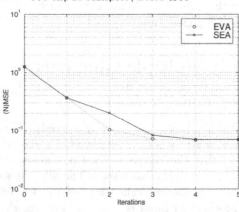

With a larger number of available samples the performance is only improved within the first iterations. As observed before, the EVA converges slightly faster than the SEA.

Conditions: QAM4 input signal with $N = 800$ samples, noise free, equaliser filter with $L + 1 = 7$ coefficients for the "difficult" channel.

Figure 2.19: MSE behaviour of EVA and SEA within the iterations

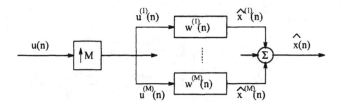

Figure 2.20: Realisation of fractionally spaced equaliser with oversampling of equaliser input.

and the respective results summed at the output, which remains symbol rate spaced. The effective equaliser now consists of a number of subsystems, with filter taps that are fractionally spaced in time. The arrangement is illustrated by figure 2.20.

Feasibility of fractional sampling requires that the transmission channel is continuous in time and that the signal possesses some excess bandwidth. To explain the latter statement, we consider a transmission medium that experience multipath propagation effects, such as a radio-frequency mobile communications environment or an underwater acoustics environment. Continuity in time allows us to take samples of the received signal at any time instant. The spacing of adjacent samples is only limited by the Nyquist rate (as given by the sampling theorem).

Increasing the sampling rate at the receiver effectively makes the fractionally spaced equaliser (FSE) a decimating filter. More samples means more computations, but the required filter length for a FSE is seen to be less than for the uniformly spaced equaliser (USE). Moreover, fractional spacing has favourable attributes in modern modem design. The sampling rate will in practice be a compromise between the requirements of the sampling theorem and the wish to keep a low computation level [23].

2.5.2 Equalisation with Multisensor Arrays

Another possibility of how to extract excess information about the unknown channel is to exploit space diversity. This can be done by employing multiple sensors at the receiver. It requires that the transmission channel is continuous is space, for example that the information signal is carried by a continuous waveform which takes multiple paths between transmitter and receiver. This means that correlated signals can be picked up by sensors that are separated slightly in space.

Figure 2.21 shows a linear array of sensors S_i that receive different ver-

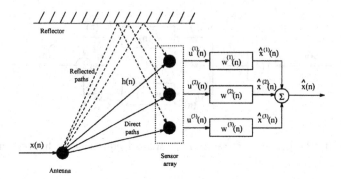

Figure 2.21: Equalisation exploiting spatial diversity using an array of $M = 3$ sensors at the receiver.

sions of the signal transmitted from the antenna A. The received signal is continuous, but we assume that the sampling devices of the sensors are synchronised and provide data sequences that are sampled at the symbol rate. The signals received by individual sensors differ because they are superimposed from different signal paths with different propagation delays.

The effect is the same as for the fractionally spaced equaliser. We obtain multiple measurements within one symbol period, though the samples are separated in space rather than in time. The equaliser input sequences from different sensors are supplied to separate subsystem equalisers, but the filter taps are not said to be fractionally spaced in a strict time sense.

2.5.3   Multichannel Signal Model

In correspondence with the requirements for fractional sampling or multisensor array application to be feasible, it will from here be assumed that the transmission channel is continuous in time and space, and that some excess bandwidth is available. So far, we have stated that fractional sampling provides us with an equaliser input sequence with some degree of diversity. The aim is to show that this provides additional information about the channel itself.

We have described two ways of acquiring more than one equaliser input sample per symbol period. The multiple samples can be seen as outputs of multiple channels that are termed virtual channels $h^{(m)}(n)$, $m = 1, \ldots, M$. When we use a multisensor array configuration, the channels are actually different in a physical sense, since the signal takes different paths from transmitter to the individual sensors. In fractionally spaced equalisation, the subchannels are divided in time. This view is the foundation of the multichannel description.

Our approach is to derive a modified signal model, where we take the consequence of the channel being a continuous bandlimited medium, characterised by $h(t)$. In this signal model, the received signal is explicitly expressed as a continuous random process. The equaliser input is still a random sequence, which is obtained by sampling of the received signal. The sampling rate is allowed to be any multiple of the baud rate. However, it should not exceed $f_s = 2f_{max}$, where f_{max} is the bandwidth of the baseband channel, since no extra information will be gained.

The whole transmission model is now rewritten with continuous time variable t. The channel input can be expressed as a continuous process carrying N symbols spaced by the symbol period T.

$$x(t) = \begin{cases} x(n), & t = 0, T, \ldots, (N-1)T \\ 0, & \text{otherwise} \end{cases} . \tag{2.90}$$

This is equivalent to

$$x(t) = \sum_{n=0}^{N-1} x(n)\delta(t - nT) . \tag{2.91}$$

$x(t)$ can be seen as the i.i.d. output $x(n)$ of a discrete data source, modulated onto a continuous carrier. The representation of $x(t)$ as an impulse train is of course purely theoretical, and will in practice be replaced by a realisable sequence of finite length pulses or rather a modulated sinusoidal carrier. From the input–output expression of the channel

$$u(t) = x(t) * h(t) = \int_{-\infty}^{+\infty} x(\xi)h(t - \xi)\,d\xi$$

we get the received signal

$$\begin{aligned} u(t) &= \int_{-\infty}^{+\infty} \left(\sum_{n=0}^{N-1} x(n)\delta(\xi - nT) \right) h(t - \xi)\,d\xi + v(t) \\ &= \sum_{k=-\infty}^{+\infty} x(kT)h(t - kT) + v(t) \end{aligned} \tag{2.92}$$

where $v(t)$ is the contribution of a continuous noise process. If N samples are taken from the continuous waveform at the baud rate, we get the equaliser input sequence

$$u(lT) = \sum_{k=-\infty}^{+\infty} x(kT)h((l-k)T) + v(lT) , \quad l = 0,\ldots,N-1 \qquad (2.93)$$

where l is a temporary discrete time index. We now want the sampling rate to exceed the baud rate, so that the symbol period T is an integer multiple of the new sampling period Δ

$$T = M \Delta$$

where M is the decimation factor. The term oversampling expresses the process that is used to obtain multiple measurements. The effect of oversampling is to replace the index l by

$$l = \frac{nM+m}{M} , \quad n = 0,\ldots,N-1 , \; m = 0,\ldots,M-1 \qquad (2.94)$$

so that the effective number of samples is NM. The oversampled sequence is given by

$$u\left(\left(n+\frac{m}{M}\right)T\right) = \sum_{k=-\infty}^{+\infty} x(kT)\, h\left(\left(n+\frac{m}{M}-k\right)T\right) + v\left(\left(n+\frac{m}{M}\right)T\right)$$

or equivalently

$$u(nT + m\Delta) = \sum_{k=-\infty}^{+\infty} x(kT)\, h((n-k)T + m\Delta) + v(nT + m\Delta) . \qquad (2.95)$$

With a more compact notation, we have

$$u_n^{(m)} = \sum_{k=-\infty}^{+\infty} x_k h_{n-k}^{(m)} + v_n^{(m)} , \quad n = 0,\ldots,N-1 , \; m = 0,\ldots,M-1 .$$
$$\qquad (2.96)$$

This equation system is called the input–output relation of the multichannel system with M virtual channels. The samples acquired during the symbol period T is seen as outputs of M separate channels $h_n^{(m)}$ with individual additive noise contributions $v_n^{(m)}$. The resulting equaliser inputs $u_n^{(m)}$ are processed separately by M different equalisers of length $L+1$. Thus, the output of the subsystem equalisers is

$$\hat{x}_n^{(m)} = \sum_{l=0}^{L} w_l^{(m)} u_{n-l}^{(m)} . \qquad (2.97)$$

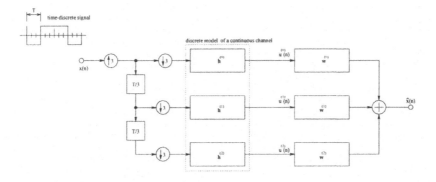

Figure 2.22: Fractional-time sampled multichannel description with $M = 3$ subchannels

After all subchannel outputs are superimposed, the symbol rate spaced equaliser output becomes

$$\hat{x}(n) = \sum_{m=1}^{M} \sum_{l=0}^{L} w_l^{(m)} u_{n-l}^{(m)} . \qquad (2.98)$$

Figure 2.22 shows the most obvious way to take a multiple sensor snapshot. The actual separation of the channels can be done in different manners. Some possible solutions to obtain a multi-dimensional channel are:

- Fractional-time sampling (see figure 2.23) for channels with excess bandwidth.

- Cyclic channel separation for cyclostationary channels (next section).

- Spatial channel separation by using space diversity.

- Multiple sensor arrangements (see figure 2.23).

- Sub-band-filtering.

2.5.4 Single Input Single Output Model

We still have not fully outlined the motivation of multichannel equalisation, except from stating that it is reasonable to exploit the information that is available in the excess bandwidth. It remains to prove that oversampled equalisers are theoretically capable of obtaining perfect equalisation with

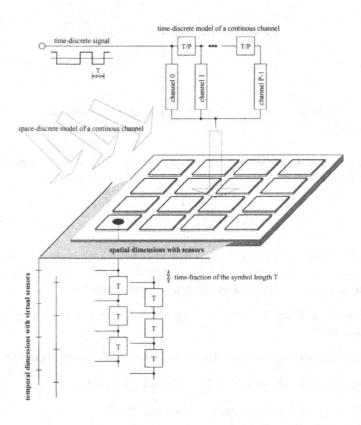

Figure 2.23: Use of multiple discrete channels in space and time domains

a finite impulse response filter. This is a powerful result with important implications for practical equaliser design.

To enable a study of the differences between symbol rate spaced and fractionally spaced equalisers, we shall deduce a matrix description of the single channel and the multiple channel signal model. Matrix and vector notations are introduced to support easier handling and allow tools of linear algebra to help us in the analysis.

First we have to assume that the channel impulse response can be modelled by a finite number of coefficients $h(n)$, $n = 0, \ldots, K$. That is, the channel is causal with finite time support. The length of the channel is apparently $(K + 1)$, while the equaliser length is $(L + 1)$. As mentioned in subsection 2.1.2, this assumption is justified in the work of Bello [1].

The uniformly spaced equaliser is described by the single input-single output relationship for the transmission channel

$$u(n) = x(n) * h(n) + v(n) = \sum_{k=0}^{K} x(n-k)h_k + v(n) \qquad (2.99)$$

where filter tap indices are subscripted for convenience. This is why it is known as the SISO model. The filtering equation for the cascaded system of channel and equaliser is the convolution

$$\hat{x}(n) = \big(x(n) * h(n) + v(n)\big) * w(n)$$
$$= \sum_{k=0}^{K}\sum_{l=0}^{L} x(n-k-l)h_k w_l + \sum_{l=0}^{L} v(n-l)w_l . \qquad (2.100)$$

If we define channel input vector $\mathbf{x}(n)$, additive noise vector $\mathbf{v}(n)$, equaliser input vector $\mathbf{u}(n)$ and equalisation filter vector $\mathbf{w}$ as

$$\mathbf{x}(n) = \begin{bmatrix} x(n) & \cdots & x(n-K-L) \end{bmatrix}^T , \qquad (2.101)$$
$$\mathbf{v}(n) = \begin{bmatrix} v(n) & \cdots & v(n-L) \end{bmatrix}^T , \qquad (2.102)$$
$$\mathbf{u}(n) = \begin{bmatrix} u(n) & \cdots & u(n-L) \end{bmatrix}^T , \qquad (2.103)$$

and

$$\mathbf{w} = \begin{bmatrix} w(0) & \cdots & w(L) \end{bmatrix}^T \qquad (2.104)$$

then a matrix formulation of the channel input-equaliser output relationship is found to be

$$\hat{x}(n) = \mathbf{w}^T \left(\mathbf{H}\mathbf{x}(n) + \mathbf{v}(n) \right) \qquad (2.105)$$

where the channel convolution matrix is defined by

$$\mathbf{H} = \begin{bmatrix} h_0 & \cdots & h_K & 0 & 0 & \cdots & 0 \\ 0 & h_0 & \cdots & h_K & 0 & \cdots & 0 \\ \vdots & & \ddots & & \ddots & & \vdots \\ 0 & \cdots & 0 & 0 & h_0 & \cdots & h_K \end{bmatrix} . \tag{2.106}$$

It exhibits a Toeplitz structure with equal elements on all diagonals. The inner brackets of Eq. (2.105) is identified as the equaliser input vector

$$\mathbf{u}(n) = \mathbf{H}\mathbf{x}(n) + \mathbf{v}(n) . \tag{2.107}$$

By the finite channel assumption, we can also write the overall impulse response consisting of cascaded channel and equaliser as

$$s(n) = h(n) * w(n) = \sum_{l=0}^{L} h(l)w(n-l) . \tag{2.108}$$

The result of linearly convolving two sequences of length $(K+1)$ and $(L+1)$ has a length of $(K+L+1)$ itself. Defining the overall system filter vector

$$\mathbf{s} = \begin{bmatrix} s_0 & \cdots & s_{K+L+1} \end{bmatrix}^T \tag{2.109}$$

we obtain the matrix representation

$$\mathbf{s} = \mathbf{H}^T \mathbf{w} . \tag{2.110}$$

2.5.5 Single Input Multiple Output Model

We have derived the following input-output relationship for the single input-multiple output transmission channel.

$$\begin{aligned} u^{(m)}(n) &= x(n) * h^{(m)}(n) + v^{(m)}(n) \\ &= \sum_{k=0}^{K} x(n-k)h_k^{(m)} + v^{(m)}(n) , \quad m = 0,\ldots,(M-1) . \end{aligned} \tag{2.111}$$

The input–output relationship between transmitted and received sequence is given by

$$\begin{aligned} \hat{x}^{(m)}(n) &= \left(x(n) * h^{(m)}(n) + v^{(m)}(n) \right) * w^{(m)}(n) \\ &= \sum_{k=0}^{K} \sum_{l=0}^{L} x(n-k-l)h_k^{(m)}w_l^{(m)} \\ &\quad + \sum_{l=0}^{L} v^{(m)}(n-l)w_l^{(m)} , \quad m = 0,\ldots,(M-1) . \end{aligned} \tag{2.112}$$

The oversampled signal model is considered in a multichannel view, as previously explained. Since this describes the channel as having a single input and multiple output, it is referred to as the SIMO model.

A matrix description of the SIMO model is found in a similar manner as for the SISO model after defining subchannel noise vector $\mathbf{v}_n^{(m)}$, subchannel equaliser input vector $\mathbf{u}_n^{(m)}$ and subchannel equalisation filter vector $\mathbf{w}^{(m)}$

$$\mathbf{v}_n^{(m)} = \left[v_n^{(m)} \quad \cdots \quad v_{n-L}^{(m)} \right]^T , \tag{2.113}$$

$$\mathbf{u}_n^{(m)} = \left[u_n^{(m)} \quad \cdots \quad u_{n-L}^{(m)} \right]^T , \tag{2.114}$$

$$\mathbf{w}^{(m)} = \left[w_0^{(m)} \quad \cdots \quad w_L^{(m)} \right]^T , \tag{2.115}$$

$$\text{with} \quad m = 0, \ldots, (M-1) .$$

The equaliser taps are now distributed over M subchannels with $L+1$ taps each, so that the effective filter length is $M(L+1)$. The subchannel filtering equation is

$$\hat{x}_n^{(m)} = \mathbf{w}^{(m)^T} \left(\mathbf{H}^{(m)} \mathbf{x}_n + \mathbf{v}_n^{(m)} \right) , \quad m = 0, \ldots, (M-1) \tag{2.116}$$

where the subchannel convolution matrix takes the form

$$\mathbf{H}^{(m)} = \begin{bmatrix} h_0^{(m)} & \cdots & h_K^{(m)} & 0 & 0 & \cdots & 0 \\ 0 & h_0^{(m)} & \cdots & h_K^{(m)} & 0 & \cdots & 0 \\ \vdots & & \ddots & & \ddots & & \vdots \\ 0 & \cdots & 0 & 0 & h_0^{(m)} & \cdots & h_K^{(m)} \end{bmatrix} . \tag{2.117}$$

Again, the subchannel equaliser input vector is found in the inner brackets of the filtering equation

$$\mathbf{u}^{(m)} = \mathbf{H}^{(m)} \mathbf{x}_n + \mathbf{v}_n^{(m)} , \quad m = 0, \ldots, (M-1) . \tag{2.118}$$

An even more compact notation is obtained by blocking matrices and vectors from the M equations in Eq. (2.117). This yields the multichannel filtering equation

$$\mathbb{U}_n = \mathbb{H}\mathbf{x}_n + \mathbf{V}_n \tag{2.119}$$

with the following block representations:

Multichannel equaliser input vector	Multichannel convolution matrix	Multichannel noise vector

$$\mathbb{U}_n = \begin{bmatrix} \mathbf{u}_n^{(0)} \\ \mathbf{u}_n^{(1)} \\ \vdots \\ \mathbf{u}_n^{(M-1)} \end{bmatrix} \qquad \mathbb{H} = \begin{bmatrix} \mathbf{H}^{(0)} \\ \mathbf{H}^{(1)} \\ \vdots \\ \mathbf{H}^{(M-1)} \end{bmatrix} \qquad \mathbf{V}_n = \begin{bmatrix} \mathbf{v}_n^{(0)} \\ \mathbf{v}_n^{(1)} \\ \vdots \\ \mathbf{v}_n^{(M-1)} \end{bmatrix} .$$

$$M(L+1) \times 1 \qquad M(L+1) \times (K+L+1) \qquad M(L+1) \times 1 \tag{2.120}$$

This result will be used later in a subspace decomposition [31]. But first we complete the treatment of the SIMO model, in analogy with the analysis of the SISO model. The equaliser output is formed from the subsystem outputs by

$$\hat{x}(n) = \sum_{m=1}^{M} \hat{x}^{(m)}(n) . \tag{2.121}$$

Thus, the overall impulse response of each subchannel is

$$\mathbf{s}^{(m)} = \mathbf{H}^{(m)^T} \mathbf{w}^{(m)} , \quad m = 0, \dots, (M-1) \tag{2.122}$$

and it follows from Eqs. (2.117) and (2.121) that the overall impulse response of the SIMO model is described by

$$\mathbf{s} = \sum_{m=0}^{M-1} \mathbf{s}^{(m)} = \mathbb{H}^T \mathbf{W} \tag{2.123}$$

where the multichannel equalisation filter vector

$$\mathbf{W} = \begin{bmatrix} \mathbf{w}^{(0)^T} & \cdots & \mathbf{w}^{(M-1)^T} \end{bmatrix}^T \tag{2.124}$$

represents the effective equaliser.

2.5.6 Full Rank Condition

Channel equalisation with a uniformly spaced FIR filter is generally impossible because of the truncated equaliser length. We will show that this problem can be bypassed with the concept of fractional sampling and multiple channel systems. Each symbol is now represented by a number of samples that offer a certain amount of diversity. The effect on the equalisation procedure is demonstrated as follows. Perfect equalisation is described by the equation

$$\mathbf{s} = \mathbb{H}^T \mathbf{W} = c\,e^{j\theta} \begin{bmatrix} 0 & \cdots & 0 & 1 & 0 & \cdots & 0 \end{bmatrix}^T . \qquad (2.125)$$

We are here concerned with the convolution matrix $\mathbb{H}$ of the SIMO model, which includes the convolution matrix of the SISO model as a special case ($M = 1$). The scaling factor $ce^{j\theta}$ and the nonzero element s_d express the permission of a gain, linear phase shift and constant delay. This makes Eq. (2.125) a sufficient and necessary equalisation criterion.

From linear algebra we know that this inhomogeneous equation has a unique solution if all columns of $\mathbb{H}$ are linearly independent, or equivalently that $\mathbb{H}$ is of full column rank. Equivalently, the whole solution space for the equaliser coefficients can be accessed by the solution

$$\mathbf{W} = \left(\mathbb{H}^T \mathbb{H} \right)^{-1} \mathbb{H}\mathbf{s} = \mathbb{H}^+ \mathbf{s} \qquad (2.126)$$

where $\mathbb{H}^+$ is defined as the pseudo-inverse of $\mathbb{H}$, if and only if $\mathbb{H}$ satisfies the full rank condition. The size of $\mathbb{H}$ is $M(L+1) \times (K+L+1)$. Thus, perfect equalisation requires that the multichannel convolution matrix satisfies

$$\text{rank}\,(\mathbb{H}) = K + L + 1 . \qquad (2.127)$$

This is the full rank condition. It implies that if

1. $M(L+1) < (K+L+1)$, we have only the trivial solution $\mathbf{W} = \mathbf{0}$.

2. $M(L+1) \geq (K+L+1)$, then non-trivial solutions may exist.

Evidently, no solution can be found for the one-dimensional SISO system. Conversely, the temporal diversity of a SIMO system may facilitate perfect equalisation, provided that Eq. (2.127) holds. The full rank condition is satisfied under three sufficient conditions given in the important convolution matrix rank theorem, derived by Tong et al. [43].

1. The transfer functions

$$H^{(m)}(z) = \sum_{k=0}^{K} h_k^{(m)} z^{-k} , \quad m = 0, \ldots, (M-1)$$

of the virtual subchannels must have no common zeros.

2. At least one of the transfer functions must be of maximum degree K.

3. The length $M(L+1)$ of the full equaliser (including taps of all subsystems) must be greater than the channel length K.

The full rank condition can be explained as follows. Denote by $\mathbf{h}_d$ the column of $\mathbb{H}$ which is associated with the nonzero element s_d of the overall system impulse response. If $\mathbf{h_d}$ is removed from $\mathbb{H}$, we get a matrix $\mathbb{H}_{hg}$ which defines the homogeneous equation system

$$\mathbb{H}_{h_g}^T \mathbf{W} = \mathbf{0} \ . \tag{2.128}$$

The perfect equalisation criterion can thus be written as

$$\begin{bmatrix} \mathbb{H}_{hg} & \mathbf{h}_d \end{bmatrix} \mathbf{W} = \begin{bmatrix} \mathbf{0} \\ e^{j\theta} \end{bmatrix} \ . \tag{2.129}$$

The size of $\mathbb{H}_{hg}$ is $M(L+1) \times (K+L)$. Solvability of the homogeneous equation will depend on the rank of $\mathbb{H}$. Specifically, it can be seen that if

1. $\operatorname{rank} \mathbb{H}_{hg} = M(L+1)$, we have the unique trivial solution $\mathbf{W} = \mathbf{0}$.

2. $\operatorname{rank} \mathbb{H}_{hg} < M(L+1)$, an infinite number of non-trivial solutions exist.

In the latter case, the system of equations is underdetermined. The nullity of $\mathbb{H}_{hg}$ is a measure of how many degrees of freedom the resulting parameterised solution possesses. Thus, if $M(L+1) = K+L+1$ we have a nullity of one.

$$\operatorname{null} \mathbb{H}_{hg} = \dim \mathbb{H}_{hg} - \operatorname{rank} \mathbb{H}_{hg} = 1 \ . \tag{2.130}$$

This means that the condition given by the inhomogeneous equation $\mathbf{h}_d^T \mathbf{W} \neq 0$ is sufficient to determine a unique solution for $\mathbf{W}$. It requires that $\mathbf{h}_d$ is linearly independent of the columns of $\mathbb{H}_{hg}$, which again makes the rank of $\mathbb{H}$ the full $(K+L+1)$ when $\mathbf{h}_d$ is inserted back in place.

A geometrical explanation of the full rank condition can be given as follows. If the number of rows $M(L+1)$ is smaller then the number of columns $(K+L+1)$, then the linear combination of the rows can only address a subspace of the whole solution space for $\mathbf{s}$, which is spanned by the column vectors of $\mathbb{H}$. Thus, it is not guaranteed that one of the desired solutions for perfect equalisation is included.

Many solutions for $\mathbf{s}$ lie outside the accessible range of the equalisation filter, which can be attributed to lack of dimensionality. The equaliser is likely to converge to points at the boundary between the accessible and the

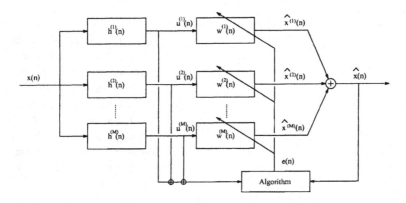

Figure 2.24: Block diagram of general fractionally spaced equaliser.

inaccessible space, since this surface may possess virtual optima or plateaus, where convergence speed slows down.

If we assume that the full rank condition of the multichannel convolution matrix is satisfied, then the homogeneous equation

$$\mathbb{H}^T_{hg(d)}\mathbf{W} = \mathbf{0} \qquad (2.131)$$

is a sufficient and necessary condition for perfect equalisation. Eq. (2.131) is the zero forcing solution, where the delay d can be chosen freely between 0 and $(K + L)$. This formulation does not give an explicit solution for $\mathbf{w}$, since the convolution matrix $\mathbb{H}_{hg(d)}$ is unknown. However, closed-form solutions can be developed from this criterion, leading to the class that is known as zero forcing algorithms.

2.5.7 Extending T-Spaced Algorithms To FSE

Fractionally spaced equalisation is commonly used to improve the performance of Bussgang and HOS-based algorithms. Successful examples include fractionally spaced CMA (FS-EVA) [10, 11, 26, 48], fractionally spaced SEA (FS-SEA) and the generalised EVA (GENEVA) [24]. As a summary, the multiple channel model maintains more information about the physical channel and increases the chances for fast converging equalisation. Whenever possible, this kind of equalisation should be applied.

Most known methods are easily extended by such a concept. Equaliser inputs are oversampled by a factor M and distributed to M equalisation filters which produce multichannel outputs. The resulting equaliser output

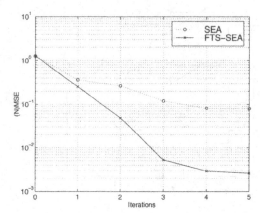

The upper graph shows the SEA with normal symbol rate sampling. In comparison, the lower graph is the fractional-time (double symbol rate) sampled algorithm for similar conditions. Obviously there is no doubt about the performance improvement.

Conditions: QAM4 input signal with $N = 300$ input samples, noise free, $P = 2$ fractionally-time sampled sub-channel equalisers with each $L + 1 = 3$ coefficients for the "difficult" channel.

Figure 2.25: MSE behaviour of fractionally-time sampled SEA within the iterations

is the mean of these, given by

$$\hat{x}(n) = \sum_{m=1}^{M} \hat{x}_n^{(m)} = \sum_{m=1}^{M} \left(\sum_{l=0}^{L} w_l^{(m)} u_{n-l}^{(m)} \right) \tag{2.132}$$

Thus, equalisation is effectively carried out by $M(L + 1)$ filter taps. The algorithm specific update procedure is used to adjust the subchannel equalisers individually, by minimising some cost function that is implicitly or explicitly based on higher-order statistics. The generalisation is demonstrated by the architecture in figure 2.24.

2.5.8   Simulations with FSE Algorithms

The superior performance properties of multiple channel equaliser is visualised in figures 2.25, 2.26, 2.27 and 2.28. For the SEA, EVA and CMA approach are the convergence behaviour of the symbol rate sampled and the fractionally-spaced sampled solution are directly compared under similar conditions. The obtained results clearly support the use of multiple systems whenever possible.

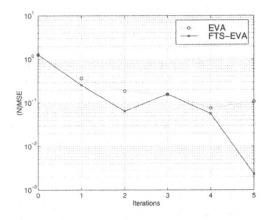

The upper graph shows the EVA with normal symbol rate sampling. The lower graph displays the behaviour of the algorithm with fractional-time sampling. Same improvements are observed as for the SEA.

Conditions: QAM4 input signal with $N = 300$ input samples, noise free, $P = 2$ fractionally-time sampled sub-channel equalisers with each $L + 1 = 3$ coefficients for the "difficult" channel.

Figure 2.26: MSE behaviour of fractionally-time sampled EVA within the iterations

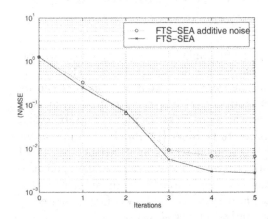

The SEA under additive noise disturbance performs equally well.

Conditions: QAM4 input signal with $N = 300$ input samples, SNR=30dB, $P = 2$ fractionally-time sampled sub-channel equalisers with each $L + 1 = 3$ coefficients for the "difficult" channel.

Figure 2.27: MSE behaviour of fractionally-time sampled SEA by additive noise disturbance

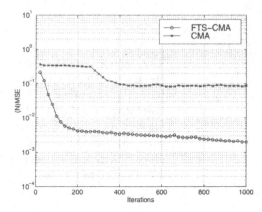

An obvious improvement is also observed in case of the CMA (without optimisation of the step-size parameter μ). The convergence behaviour is clearly better for the multi-channel case.

Conditions: QAM4 input signal, noise free, up to $N = 1000$ input samples adaptively processed by an gradient search algorithm, step size $\mu = 10^{-5}$, $P = 2$ fractionally-time sampled sub-channel equalisers with each $L + 1 = 3$ coefficient and symbol rate sampled with $L + 1 = 7$ coefficients for the "difficult" channel.

Figure 2.28: Comparison of CMA and fractionally-time sampled CMA

2.6 Algorithms Based on Cyclostationary Statistics

Identification and inverse modelling of nonminimum phase systems with stationary inputs require methods that involve at least third-order statistics to extract phase information about the system. In contrast, the task can be accomplished with second-order statistics (SOS) for a certain class of nonstationary signals. This is true only for signals that exhibit cyclostationarity.

This discovery has been tributed to Gardner [14–16]. It is an important revelation, since many signals in communications applications are indeed cyclostationary. Consequently, the concept has been successfully applied in various problems such as detection, filtering, parameter estimation, direction finding and identification of nonlinear systems [44] It was first exploited in blind identification and equalisation by Tong et al. [43].

Stationarity of a stochastic process requires that its statistics are constant in time. Cyclostationarity, on the other hand, implies that signal statistics are periodic [13,17]. Cyclostationary signals are periodically correlated, thus possessing some spectral diversity that be exploited in identification and equalisation of non-minimum phase systems. As a counterpart to wide sense stationarity (wss), a stochastic process is wide sense cyclostationary if it satisfies

$$E[x(t + T)] = E[x(t)] \qquad (2.133)$$

and

$$\begin{aligned} R_{xx}(t_1, t_2) &= E[x(t_1)x^*(t_2)] \\ &= E[x(t_1 + T)x^*(t_2 + T)] \qquad (2.134) \\ &= R_{xx}(t_1 + T, t_2 + T) \end{aligned}$$

where $R_{xx}(\cdot)$ is the autocorrelation function of $x(t)$. The constraints for a discrete-time random sequence are similar, replacing continuous time variables with discrete time indices.

Continuous phase and amplitude modulated signals are among those exhibiting cyclostationarity. This includes both the transmitted signal and the distorted received signal, as will be seen in the sequel after we have adopted a continuous signal model. When the received signal is sampled at the receiver, cyclostationarity of the resulting sequence $u(n)$ will depend on the sampling rate. Specifically, the discrete equaliser input will be cyclostationary only if the received signal is oversampled with respect to the symbol rate at the transmitter.

This takes us back to the fundamentals of fractional sampling, which was discussed in a general context in the previous section, without targeting algorithms based on cyclostationary statistics (CS) in particular. In the sequel, we will discuss the motivation and achievements of methods exploiting second-order cyclostationarity. We will then present some explicit algorithms.

Fractional sampling of the received signal will enforce cyclostationarity on the discrete equaliser input, but is feasible only if the received signal has some excess bandwidth, so that decimation does not violate the sampling theorem. However, cyclostationarity can be induced on the discrete received sequence without resorting to oversampling. The approach is known as modulation induced cyclostationarity, where cyclostationarity is introduced already at the transmitter by means of an encoder. It is also termed forced cyclostationarity, since signal characteristics are enforced by modulation rather than exploiting inherent signal properties.

Different encoding schemes have been suggested to obtain modulation induced cyclostationarity. Chevreuil and Loubaton [7] modulate the input stream by a deterministic and almost periodic sequence. An alternative approach is reported independently by Tsatsanis and Giannakis [18, 47], who use precoding filterbanks or repetition codes to obtain modulation with a strictly periodic sequence. It has been shown that equalisation algorithms can be developed from these approaches that have no restrictions on the channel zeros and the colour of the additive noise [37]. Nevertheless, we will concentrate on cyclostationarity introduced by fractional sampling.

2.6.1 Cyclostationarity of Modulated Input

If we return to the continuous signal model that was introduced for the treatment of fractionally spaced equalisation, then we have the continuous channel input

$$x(t) = \sum_{n=0}^{N-1} x(n)\delta(t - nT) \, . \tag{2.135}$$

It is now easily seen that $x(t)$ is cyclostationary, since it can be derived that

$$E[x(t_1)x^*(t_2)] = \begin{cases} \sigma_x^2 \delta(t_1 - t_2), & t_1, t_2 = 0, T, \dots, (N-1)T \, , \\ 0 \, , & \text{otherwise} \end{cases} \tag{2.136}$$

$$= E[x(t_1 + T)x^*(t_2 + T)] \, .$$

This result must again be taken for demonstration purposes only, since an impulse train representation of the channel input is not realisable in practise. The actual form which $R_{xx}(t_1, t_2)$ takes in depends on what modulation scheme is used. If the input data stream is modulated onto a continuous carrier using phase or amplitude modulation, or a combination of those (e.g. QAM), then it is readily shown that the continuous input is second-order cyclostationary.

We proceed by regarding the channel as a continuous bandlimited medium described by $h(t)$. From Eq. (2.136), the equaliser input becomes

$$u(t) = \int_{-\infty}^{+\infty} \left(\sum_{n=0}^{N-1} x(\tau)\delta(\tau - nT) \right) h(t - \tau) \, d\tau + v(t)$$

$$= \sum_{k=-\infty}^{+\infty} x(kT)h(t - kT) + v(t) \tag{2.137}$$

where $v(t)$ is the contribution of a continuous noise process. Under the assumption that $v(t)$ is i.i.d., wss and independent of $x(t)$, it follows that the received signal $u(t)$ is also wide sense cyclostationary. The mean of $u(t)$ is constant (periodic with any period), while the autocorrelation function $R_{uu}(\tau)$ is a periodic function of the symbol period T.

A discrete sequence that can be processed digitally is obtained by sampling $u(t)$. If samples are taken at the baud rate $1/T$, then the resulting $u(n)$ will be wss. On the other hand, sampling $u(t)$ at an integer multiple of the baud rate will produce a wide sense cyclostationary equaliser input sequence.

2.6.2 Spectral Diversity

To understand how wide sense cyclostationarity can be exploited in identification of nonminimum phase systems, we shall analyse some second-order signal statistics in the frequency domain [44,45]. A wide sense stationary sequence has the property that its autocorrelation function is only dependent on the discrete time difference $\tau = (n_2 - n_1)T$ between the correlated signals. Furthermore, from Wiener-Khinchin theorem we know that the power spectrum is the Fourier transform of the autocorrelation function. Thus, if $u(n)$ is wss, we have

$$C_{2u}(\omega) = C_{2x}(\omega)H(\omega)H^*(\omega) = C_{2x}(\omega)|H(\omega)|^2 \qquad (2.138)$$

where $C_{2u}(\omega)$ and $C_{2x}(\omega)$ are the respective power (second-order cumulant) spectra of $u(n)$ and $x(n)$, and $H(\omega)$ is the frequency response of $h(n)$. From this expression it is clear that we can only identify the magnitude response from the second-order spectrum. If the system is nonminimum phase, then the phase response is not uniquely determined. In contrast, the autocorrelation function of a wide sense cyclostationary sequence $u(n)$ depends on exact time (specific sample indices n_1 and n_2). Therefore, the Fourier transform becomes two-dimensional and gives the power spectrum

$$C_{2u}(\omega, \nu) = C_{2x}(\omega, \nu)H(\omega)H^*(\nu) \qquad (2.139)$$

from which both the magnitude response and the phase response of the channel can be identified without ambiguity.

CS-based equalisation algorithms have attractive features, since they rely on second-order statistics only. The number of samples needed to obtain empirical statistics is known to be almost exponential to the order of the statistics [4]. Thus, CS-based algorithms require less data to obtain good estimates of the inverse channel.

Convergence rates are superior to Bussgang and HOS-based algorithms. Hence, CS-based algorithms are more capable of tracking faster changes in the context of time-varying channels. Theoretically, they provide perfect equalisation, given the true correlation function of the process. The exact solution is approached asymptotically by use of empirical functions.

Furthermore, there are no restrictions imposed on the distribution of the input sequence. CS-based algorithms can equalise channels with Gaussian inputs. On the other hand, HOS-based algorithms have the feature of discarding effects of Gaussian noise.

2.6.3   Overview of CS-Based Algorithms

Equalisation algorithms based on second-order cyclostationary statistics exploit the matrix structure derived under the SIMO model to formulate an explicit equalisation scheme. The first methods presented were directed towards channel identification, rather than direct equalisation. These methods rely on subspace separability of signal and noise. Such approaches are developed by Tong et al. [44,45], Xu et al. [50] and Moulines et al. [31].

Subspace decomposition in blind equalisation requires knowledge of the channel order. A weakness of the subspace methods is that they tend to be sensitive to errors in channel order estimates. An alternative approach based on linear prediction theory is described in papers by Slock [41,42] and Abed Meriam et al. [30]. This method is more robust to overestimation of the channel order. A recent attempt to overcome the channel order estimation problem is made by Ding et al. [9]. This algorithm does not go through the intermediate step of channel identification.

Another interesting approach is the methods relying on modulation induced cyclostationarity [7,18,47]. It has been shown that these methods can can offer unique identification of SISO channels irrespective of the position of channel zeros, colour of additive stationary noise and channel order overestimation error, provided that the period of cyclostationarity which is enforced at by encoding at the channel input, is greater than half the channel length [37].

Some explicit algorithms based on cyclostationary statistics are described in the following. The zero forcing principle described in subsection 2.5.6 can be implemented in different ways. One such approach is presented below. To further demonstrate how channel characteristics can be explicitly extracted from second-order autocorrelation functions, we will also consider the subspace decomposition which is the foundation of Moulines algorithm [31].

2.6.4   Zero Forcing Algorithm

A closed-form solution based on the zero-forcing solution in Eq. (2.131) is given in the following. Let $\mathbf{w}^{(o)}$ be the vector representation of the perfect equaliser $w^{(0)}(n)$ with an undelayed identity operation as cascaded impulse response. That is, $w^{(0)}(n)$ is chosen so that

$$s(n) = h(n)w^{(0)}(n) = e^{j\theta}\delta(n) \ . \tag{2.140}$$

Correspondingly, we write the filter vector that obtains perfect equalisation with delay d as $\mathbf{w}^{(d)}$. It corresponds to the equaliser $w^{(d)}(n)$ which obtains

$$s(n) = h(n) * w^{(d)}(n) = e^{j\theta}\delta(n - d). \tag{2.141}$$

Calculation of an output $\hat{x}(n)$ requires that $L+1$ symbols are received. The received signal is oversampled M times such that the equaliser input is an input sequence of $M(L+1)$ samples. It is written on vector form as

$$\mathbf{u}(n) = \left[u_n^{(1)} \quad \cdots \quad u_n^{(M)} \quad \cdots \quad u_{n-L}^{(1)} \quad \cdots \quad u_{n-L}^{(M)} \right]^T , \quad n = L+1, \ldots, N \tag{2.142}$$

which is processed by the multichannel equaliser vector

$$\mathbf{w} = \left[w_0^{(1)} \quad \cdots \quad w_0^{(M)} \quad \cdots \quad w_L^{(1)} \quad \cdots \quad w_L^{(M)} \right]^T . \tag{2.143}$$

The motivation of the algorithm is the observation that there are different ways of producing the estimate $\hat{x}(n-d)$. The output $\hat{x}(n-d)$ will result both when $\mathbf{u}(n)$ is filtered by $\mathbf{w}^{(d)}$ and when $\mathbf{u}(n-d)$ is filtered by $\mathbf{w}^{(0)}$, as well as other combinations that might exist. We want to use the identity

$$\mathbf{u}^T(n)\mathbf{w}^{(d)} = \mathbf{u}^T(n-d)\,\mathbf{w}^{(0)} \tag{2.144}$$

or equivalently

$$\mathbf{u}^T(n)\mathbf{w}^{(d)} - \mathbf{u}^T(n-d)\,\mathbf{w}^{(0)} = \mathbf{0} \tag{2.145}$$

to determine $\mathbf{w}$. This reasoning can be extended to all $\hat{x}(n)$ for $n = (L+1), \ldots, (N-d)$. We thus have the equation system

$$\begin{bmatrix} \mathbf{u}^T(N) \\ \vdots \\ \mathbf{u}^T(L+d+1) \end{bmatrix} \mathbf{w}^{(d)} - \begin{pmatrix} \mathbf{u}^T(N-d) \\ \vdots \\ \mathbf{u}^T(L+1) \end{pmatrix} \mathbf{w}^{(0)} = \mathbf{0} . \tag{2.146}$$

The matrix equation is rewritten adopting the following notation

$$\mathbf{U}^{(0)}\mathbf{w}^{(d)} - \mathbf{U}^{(d)}\mathbf{w}^{(0)} = \mathbf{0} . \tag{2.147}$$

This can be further simplified by defining

$$\mathbf{U}\tilde{\mathbf{w}} = \begin{bmatrix} -\mathbf{U}^{(d)} & \mathbf{U}^{(0)} \end{bmatrix} \begin{bmatrix} \mathbf{w}^{(0)} \\ \mathbf{w}^{(d)} \end{bmatrix} = \mathbf{0} . \tag{2.148}$$

Under practical constraints it is impossible to compute a solution for $\tilde{\mathbf{w}}$ that gives the exact null vector in Eq. (2.148). Therefore, the null vector is replaced by the error vector $\boldsymbol{\epsilon}$. We then minimise the norm of $\boldsymbol{\epsilon}$, which is evaluated from

$$\begin{aligned} \|\boldsymbol{\epsilon}\|^2 &= \boldsymbol{\epsilon}^{*T}\boldsymbol{\epsilon} \\ &= (\mathbf{U}\tilde{\mathbf{w}})^{*T}(\mathbf{U}\tilde{\mathbf{w}}) \\ &= \tilde{\mathbf{w}}^{*T}(\mathbf{U}^{*T}\mathbf{U})\tilde{\mathbf{w}} . \end{aligned} \tag{2.149}$$

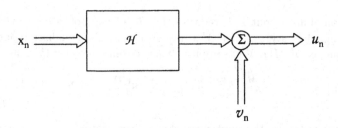

Figure 2.29: Compact matrix representation of oversampled multichannel model.

This is seen to be an eigenvector equation, which can be reformulated in the familiar form

$$\left(\mathbf{U}^{*T}\mathbf{U}\right)\tilde{\mathbf{w}} = \lambda\tilde{\mathbf{w}} \tag{2.150}$$

with eigenvalue $\lambda = \|\epsilon\|^2$. Thus, it is obvious that the norm of the error vector is minimised by choosing the eigenvector that is associated with the smallest eigenvalue $\lambda_{\min}$. This is the mean squared error solution for the equaliser. The MSE can be determined from

$$\mathrm{MSE} = \frac{\lambda_{\min}}{N - L - d}. \tag{2.151}$$

The solution $\tilde{\mathbf{w}}$ contains the equalisers $\mathbf{w}^{(0)}$ and $\mathbf{w}^{(d)}$. The disadvantages of the zero-forcing algorithm lies in the requirements of the filtering matrix theorem. The rank of $\mathbf{U}^{*T}\mathbf{U}$ is unknown, and depends very much on the channel and the parameters selected, such as oversampling factor M and equaliser subsystem length $(L + 1)$.

If $\mathbf{U}^{*T}\mathbf{U}$ is not full rank, there will be eigenvalues that are equal to zero. Hence, eigenvectors corresponding to larger eigenvalues must be chosen as estimates of $\tilde{\mathbf{w}}$. Consequently, the error in ϵ is greater and equalisation performance is reduced. This explains why the algorithm has not yet found any practical use.

2.6.5   Subspace Decomposition

The starting point of the subspace decomposition is the matrix formulation of the multichannel filtering equation

$$\mathbb{U}_n = \mathbb{H}\mathbf{x}_n + \mathbf{V}_n. \tag{2.152}$$

A corresponding compact block diagram of the oversampled channel is shown in figure 2.29. We assume that the entries of the signal vector $\mathbf{x}_n$ and multi-channel noise vector $\mathbf{V}_n$ are uncorrelated, and that the multichannel convolution matrix satisfies the full rank condition. In the search for an explicit algorithm, we examine the correlation matrices of the elements in Eq. (2.152). The transmitted signal vector is zero mean and has correlation matrix

$$\mathbf{R}_x = \mathrm{E}\left[\mathbf{x}_n\mathbf{x}_n^{*T}\right] . \tag{2.153}$$

The multichannel noise vector is zero mean with correlation matrix

$$\mathbf{R}_v = \mathrm{E}\left[\mathbf{V}_n\mathbf{V}_n^{*T}\right] = \sigma_v^2\mathbf{I} , \tag{2.154}$$

where the noise variance σ_v^2 is unknown, and $\mathbb{I}$ is an identity matrix. The received multichannel signal vector is zero mean with correlation matrix

$$
\begin{aligned}
\mathbf{R}_u &= \mathrm{E}\left[\mathbb{U}_n\mathbb{U}_n^{*T}\right] \\
&= \mathrm{E}\left[\mathbb{H}\mathbf{x}_n + \mathbf{V}_n)(\mathbb{H}\mathbf{x}_n + \mathbf{V}_n)^{*T}\right] \\
&= \mathrm{E}\left[\mathbb{H}\mathbf{x}_n\mathbf{x}_n^{*T}\mathbb{H}^{*T}\right] + \mathrm{E}\left[\mathbf{V}_n\mathbf{V}_n^{*T}\right] \\
&= \mathbb{H}\mathbf{R}_x\mathbb{H}^{*T} + \mathbf{R}_v .
\end{aligned}
\tag{2.155}
$$

In the following we make use of some results from linear algebra. Mercer's theorem [21] states that a correlation matrix can be decomposed in terms of its eigenvectors and eigenvalues, by means of the unitary similarity transformation. For the correlation matrix of the multichannel received signal vector

$$\mathbf{R}_u = \sum_{i=1}^{M(L+1)} \lambda_i\mathbf{q}_i\mathbf{q}_i^{*T} \tag{2.156}$$

where $\mathbf{q}_i$ is the ith eigenvector of $\mathbf{R}_u$ with associated eigenvalue λ_i. The dimension and rank of the received signal correlation matrix is $M(L+1)$. Next important result is that an eigenvalue decomposition of $\mathbf{R}_u$ gives

$$
\begin{aligned}
\mathbf{R}_u &= \mathbf{Q}\,\mathrm{diag}\left(\left[\lambda_1, \cdots \lambda_{K+L+1}\ 0\ \cdots\ 0\right]\right)\mathbf{Q}^{*T} \\
&\quad + \mathbf{Q}\,\mathrm{diag}\left(\left[\sigma_v^2 \cdots \sigma_v^2\right]\right)\mathbf{Q}^{*T} \\
&= \mathbf{Q}\,\mathrm{diag}\left(\left[\lambda_1 + \sigma_v^2 \cdots \lambda_{K+L+1} + \sigma_v^2\ \sigma_v^2 \cdots \sigma_v^2\right]\right)\mathbf{Q}^{*T} .
\end{aligned}
\tag{2.157}
$$

Here $\mathbf{Q}$ denotes the unitary diagonalising matrix and $\mathrm{diag}\left((\cdot)\right)$ is a diagonal matrix where the entries on the diagonal are given in descending order. The diagonal elements are eigenvalues of $\mathbf{R}_u$. These can be decomposed as a sum of eigenvalues of $\mathbb{H}\mathbf{R}_x\mathbb{H}^{*T}$ and $\mathbf{R}_v$, as seen from Eq. (2.155).

From the full rank assumption of $\mathbb{H}$, it is evident that $\mathbb{H}\mathbf{R}_x\mathbb{H}^{*T}$ has $(K + L + 1)$ non-zero eigenvalues. In theory, σ_v^2 and K can be obtained by determining the most significant eigenvalues $\mathbf{R}_u$. In practice, a threshold test can be employed to extract the parameters from an empirical covariance matrix. The task is not trivial, however, and again reflects the problem of channel order estimation.

Let the eigenvectors of $\mathbf{R}_u$, as seen in the decomposition of Eq. (2.156), be sorted according to descending magnitude of corresponding eigenvalues λ_i. We further define the partition

$$\{\mathbf{q}_i\}_{k=1}^{M(L+1)} = \{\mathbf{f}_1, \ldots, \mathbf{f}_{K+L+1}, \mathbf{g}_{K+L+2}, \ldots, \mathbf{g}_{M(L+1)}\} . \qquad (2.158)$$

It can then be shown that the eigenvectors $\{\mathbf{f}_i\}$ span the signal subspace $\mathcal{S}$, while the eigenvectors $\{\mathbf{g}_i\}$ span the noise subspace $\mathcal{N}$. We can express this as

$$\mathbf{R}_u = \sum_{k=1}^{K+L+1} \lambda_k \mathbf{f}_i \mathbf{f}_i^{*T} + \sum_{k=K+L+1}^{M(L+1)} \lambda_k \mathbf{g}_i \mathbf{g}_i^{*T} \qquad (2.159)$$

which is equivalent to Eq. (2.156). Subspaces $\mathcal{S}$ and $\mathcal{N}$ are orthogonal compliments. From the orthogonal property, we have

$$\begin{aligned}
\mathbf{R}_u \mathbf{g}_i &= (\mathbb{H}\mathbf{R}_x \mathbb{H}^{*T})\mathbf{g}_i + \mathbf{R}_v \mathbf{g}_i \\
&= 0 + \sigma_v^2 \mathbf{g}_i \\
&= \sigma_v^2 \mathbf{g}_i, \quad i = (K + L + 2), \ldots, M(L + 1) .
\end{aligned} \qquad (2.160)$$

Specifically, we observe that

$$\mathbb{H}\mathbf{R}_x \mathbb{H}^{*T} \mathbf{g}_i = 0 , \quad i = (K + L + 2), \ldots, M(L + 1) . \qquad (2.161)$$

Both $\mathbb{H}$ and $\mathbf{R}_x$ are full rank matrices, $\mathbb{H}$ by assumption of solvability and $\mathbf{R}_x$ by definition. Thus, the multichannel convolution matrix, and hence the unknown channel, can be identified from

$$\mathbb{H}\mathbf{g}_i = 0 , \quad i = (L + L + 2), \ldots, M(L + 1) . \qquad (2.162)$$

The columns of the unknown multichannel convolution matrix are orthogonal with the eigenvectors that span the noise subspace $\mathcal{N}$. By further analysis, closed-form solutions for the unknown channel and eventually the sought equaliser can be determined [31].

2.7 General Convergence Considerations

2.7.1 Gradient Descent Algorithms

Adaptive algorithms realised with gradient search techniques of the form

$$\mathbf{w}(n) = \mathbf{w}(n-1) - \mu\left(\nabla_{\mathbf{w}}J\right)^* \tag{2.163}$$

converge when the gradient term equals zero. A general cost function min-imises the distance between the desired signal or its estimation $\hat{d}(n)$ (output of nonlinear estimator) and the output signal $\hat{x}(n)$. This can be obtained by mean square error cost function in the following notation

$$
\begin{aligned}
J(n) &= \mathrm{E}\left[\left|\hat{d}(n) - \hat{x}(n)\right|^2\right] \\
&= \mathrm{E}\left[\left(\hat{d}(n) - \mathbf{w}^T\mathbf{u}\right)\left(\hat{d}(n) - \mathbf{w}^T\mathbf{u}\right)^{*T}\right] \\
&= \mathrm{E}\left[|\hat{d}(n)|^2 - \hat{d}(n)\mathbf{w}^*\mathbf{u}^{*T}(n) - \mathbf{w}^T\mathbf{u}(n)\hat{d}^*(n) + \mathbf{w}^T\mathbf{u}(n)\mathbf{u}^{*T}(n)\mathbf{w}^*\right] .
\end{aligned}
$$

Differentiation of the cost function with respect to $\mathbf{w}$ proceeds as follows

$$
\begin{aligned}
\nabla_{\mathbf{w}}J(n) &= \mathrm{E}\left[\mathbf{u}(n)\underbrace{\mathbf{u}^{*T}(n)\mathbf{w}^*}_{\hat{x}^*(n)} - \mathbf{u}(n)\hat{d}^*(n)\right] \\
&= \mathrm{E}\left[\mathbf{u}(n)\left(\hat{x}(n) - \hat{d}(n)\right)^*\right] \\
&= \mathrm{E}[\mathbf{u}(n)e^*(n)] .
\end{aligned}
$$

After left-multiplying with the transposed equaliser filter vector $\mathbf{w}^T$ the following equilibrium is required

$$\mathrm{E}[\hat{x}(n)\hat{x}^*(n)] = \mathrm{E}\left[\hat{x}(n)\hat{d}^*(n)\right] . \tag{2.164}$$

The application of a nonlinear estimator $\hat{d}(n) = g(\hat{x}(n))$ requires special properties of this nonlinearity. Processes which satisfy the equilibrium in Eq. (2.164), such as Sato and Godard and other approaches previously presented, are called Bussgang.

2.7.2 Algorithm and Channel Dependent Equilibria

The final convergence state is observed for

$$\left(\nabla_{\mathbf{w}}J\right)^* = \mathrm{E}\left[\mathbf{u}^*(n)e(\hat{x}(n))\right] = \mathbf{0} .$$

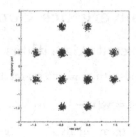

Figure 2.30: Possible appearance of ADEs for an i.i.d. 4-ary-QAM input signal (here for the Sato algorithm)

For the noiseless case, the last equation can be expressed as

$$\mathbb{H}^* E\big[\mathbf{x}^*(n)e\big(\mathbf{W}^T \mathbb{H}\mathbf{x}(n)\big)\big] = \mathbf{0}$$

where (as defined earlier and rewritten for convenience)

$$\mathbb{U} = \mathbb{H}\mathbf{x}(n) , \qquad \text{and} \qquad \hat{x}(n) = \mathbf{W}^T \mathbb{H}\mathbf{x}(n) .$$

Referring to [8], the trivial solutions

$$E\big[\mathbf{x}^*(n)e\big(\mathbf{W}^T \mathbb{H}\mathbf{x}(n)\big)\big] = \mathbf{0}$$

are named the algorithm dependent equilibria (ADE) (see figure 2.30) and the nontrivial solutions

$$E\big[\mathbf{x}^*(n)e\big(\mathbf{W}^T \mathbb{H}\mathbf{x}(n)\big)\big] \neq \mathbf{0}$$

are named the channel dependent equilibria (CDEs), because of their dependence on the channel. CDEs only appear in the multiple channel case, if the channel matrix contains a nontrivial null-space

$$\dim\big(\,\text{null}\,(\mathbb{H})\big) = M(L+1) - \text{rank}\,(\mathbf{H}) > 0 , \qquad (2.165)$$

which is possible for $P(L+1) > (K+L+1)$. Therefore excessive oversampled signals or an outsized equaliser filter length can cause CDEs that prevent channel equalisation (See examples in figure 2.31). However, some other cost function might avoid these equilibria.

To suppress the existence of a null-space it is sufficient that

a) the sub-channels polynoms with the coefficients $\big\{\mathbf{h}^{(m)}\big\}_{m=1}^{M}$ must not have common roots and

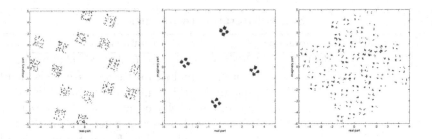

Figure 2.31: Possible appearances of CDEs without or with insufficient channel equalisation for an 4-ary-QAM input signal $(\dim(\mathrm{null}(\mathbb{H})) = 3)$

b) from the full rank condition, the equaliser filter length should be

$$(L + 1) \simeq \frac{K}{(M - 1)} \; .$$

The requirements (a) and (b) together are mentioned as the length and zero condition in literature. Note that a null-space and therefore CDEs cannot appear in the single channel case. Therefore the analysis of the existence of ADEs is simpler if carried out as analysis in a single channel environment.

2.8 Discussion

The major advantages of blind equalisation are the absence of the need for training and the consequent increase in bandwidth. In blind equalisation one essentially identifies the inverse of an unknown linear time-invariant (possibly non-minimum phase) system - both its magnitude and its phase. Magnitude component can be identified using second order statistics of the output but higher order statistics allow the identification of the phase component as well. Many types of algorithms have been considered in this chapter. The Bussgang type algorithms use higher order statistics implicitly, whereas the superexponential type algorithms use higher order statistics explicitly. It is shown that the performance of the superexponential type algorithms, for small number of samples, is superior to that of the Bussgang type algorithms. Latest advances show that some modifications of the superexponential type algorithms (for example, EVA) have even better performance [22].

Symbol-iterative techniques		
Method	Global convergence	Comment
DD	NO	Multiple ADEs are probable. This method is usually applied after adequate equalisation quality is achieved, to reduce the MSE in a maximum likelihood approach (chooses the closest symbol from a given alphabet). In this case, undesired ADEs will hardly appear because the gradient search is restricted to a small vector space.
Sato	NO	In general, applications are not globally convergent and also tend to converge to non-global minima. Under MCC, for sub-Gaussian and and uniformly distributed inputs, the algorithm always converges to the global ADE [29].
CMA	YES	QAM: No local ADE exists for MCC. A proof considering QAM inputs is found in [10,12,27,38]. The CMA can be seen as first-order stochastic approximation of the SEA algorithm.
BG	NO	Local minima exists [29], see also Sato
SG	NO	Local minima exists [29]
Batch techniques		
ZF	—	CDEs are most likely caused by underestimation of the channel order.
TEA	?	
SEA	YES	No local ADEs [10,27,38,39].
EVA	YES	No local ADEs. This algorithm is a variation of the SEA and basically exhibits the same performance.

Abbreviations: DD: decision directed, QAM: quadrature amplitude modulation, MCC: multiple channel condition, ADE: algorithm depending equilibria, CDE: channel depending equilibria, CMA: constant modulus algorithm, BG: Benveniste-Goursat algorithm, SG stop-and-go algorithm, ZF: zero forcing algorithm, SEA: super-exponential algorithm, EVA: eigenvector algorithm, TEA: tricepstrum equalisation algorithm

Table 2.2: ADE convergence statements of blind equalisation methods for discrete i.i.d. signals

References

[1] P. A. Bello. Characterization of randomly time-invariant linear channels. *IEEE Transactions on Communications Systems*, CS-11:360–393, 1963.

[2] A. Benveniste and M. Goursat. Blind equalizers. *IEEE Transactions on Communications*, COM-32:871–883, 1984.

[3] A. Benveniste and G. Ruget. Robust identification of a nonminimum phase system: blind adjustment of a linear equalizer in data communications. *IEEE Transactions on Automated Control*, pages 385–399, 1980.

[4] D. R. Brillinger. *Time Series: Data Analysis and Theory*. Holt, Rinehart and Winston, 1974.

[5] J. J. Bussgang. Crosscorrelation functions of amplitude-distorted Gaussian signals. Technical Report 216, M.I.T., 1952.

[6] Y. Chen, C. L. Nikias, and J. G. Proakis. Crimno: Criterion with memory nonlinearity for blind equalization. In *Proceedings of International Signal Processing Workshop on Higher-Order Statistics*, pages 57–90, 1991.

[7] A. Chevreutil and P. Loubaton. Blind second-order identification of second-order FIR channels: Forced cyclo-stationarity and structured subspace method. *IEEE Signal Processing Letters*, 4:204–206, 1997.

[8] Z. Ding. On convergence analysis of fractionally spaced adaptive blind equalizers. *IEEE Transactions Signal Processing*, 1997.

[9] Z. Ding, I. B. Collings, and R. W. Liu. A new blind zeroforcing equalizer for multichannel systems. In *Proceedings of ICASSP'98*, Seattle, Washington, 1998.

[10] I. Fijalkow, F. L. de Victoria, and C. R. Johnson. Fractionaly spaced equalization with CMA. In *Digital Signal Processing Workshop*, Yosemite, California, 1994.

[11] I. Fijalkow, C. E. Manlove, and C. R. Johnson. Adaptive fractionally spaced blind CMA equalization: Excess MSE. *IEEE Transactions on Signal Processing*, SP-46:227–231, 1998.

[12] G. J. Foschini. Equalizing without altering and detecting data. *Bell Systems Technical Journal*, 64:1885–1911, 1985.

[13] L. E. Franks. *Signal Theory*. Prentice Hall, Englewood Cliffs, New Jersey, 1969.

[14] W. A. Gardner. *Introduction to Random Processes with Applications to Signals and Systems*. Macmillan, 1985.

[15] W. A. Gardner. Exploiting spectral redundancy in cyclostationary signals. *IEEE Signal Processing Magazine*, 8(2):14–36, 1991.

[16] W. A. Gardner. A new method of channel identification. *IEEE Transactions on Communication*, COM-39:813–817, 1991.

[17] W. A. Gardner and L. E. Franks. Characterization of cyclostationary random signal processes. *IEEE Transactions on Information Theory*, IT-21:4–14, 1975.

[18] G. B. Giannakis. Filterbanks for blind channel identification and equalization. *IEEE Signal Processing Letters*, 4:184–189, 1997.

[19] D. N. Godard. Self recovering equalization and carrier tracking in two-dimensional data communication systems. *IEEE Transactions on Communications*, COM-28:1867–1875, 1980.

[20] D. Hatzinakos and C. L. Nikias. Blind equalization using a tricepstrum based algorithm. *IEEE Transactions on Communications*, COM-39:669–682, 1991.

[21] S. Haykin. *Adaptive Filter Theory*. Prentice-Hall, Upper Saddle River, New Jersey, 3 edition, 1996.

[22] F. Herrmann and A. K. Nandi. Low computation blind super-exponential equaliser. Accepted for publication in IEE Electronics Letters, 1998.

[23] I. F. J R Treichler and C. R. J. Jr. Fractionally spaced equalizers. *IEEE Signal Processing Magazine*, 13(3):65–81, 1996.

[24] B. Jellonek, D. Boss, and K. D. Kammeyer. Generalized eigenvector algorithm for blind equalization. *EURASIP Signal Processing*, pages 237–264, 1997.

[25] B. Jellonek and K. D. Kammeyer. A closed-form solution to blind equalization. *EURASIP Signal Processing*, 36:251–259, 1994.

[26] J. F. Leblanc, I. Fijalkow, and C. R. Johnson. Blind adaptive fractionally spaced CMA equalizer error surface characterizations: Effect of source distributions. In *IEEE Proceedings of ICASSP'96*, Princeton, New Jersey, 1996.

[27] Y. Li and Z. Ding. Convergence analysis of finite length blind adaptive equalizers. *IEEE Transactions on Signal Processing*, SP-43:2120–2129, 1995.

[28] O. Macchi and E. Eweda. Convergence analysis of self-adaptive equalizers. *IEEE Transactions on Information Theory*, IT-30:161–176, 1984.

[29] J. E. Mazo. Analysis of decision-directed equalizer convergence. *Bell Systems Technical Journal*, 59:1857–1876, 1980.

[30] K. A. Meriam, P. Duhamel, J. F. Cardoso, L. Loubaton, S. Mayrargue, E. Moulines, and D. Slock. Prediction error methods for time-domain blind identification of multichannel FIR filters. In *Proceedings of ICASSP'95*, pages 1968–1971, Detroit, Michigan, 1995.

[31] E. Moulines, J. F. Cardoso, and S. Mayrargue. Subspace methods for blind identification of multichannel FIR filters. *IEEE Transactions on Signal Processing*, SP-43:516–525, 1995.

[32] R. Pan and C. R. Nikias. The complex cepstrum of higher order cumulants and nonminimum phase identification. *IEEE Transactions on Acoustics and Speech Signal Processing*, ASSP-36:186–205, 1988.

[33] G. Picchi and G. Prati. Blind equalization and carrier recovery using a 'stop-and-go' decision-directed algorithm. *IEEE Transactions on Communications*, COM-35:877–887, 1987.

[34] B. Porat and B. Friedlander. Blind equalization of digital communications channels using higher-order moments. *IEEE Transactions on Signal Processing*, SP-39:522–526, 1991.

[35] J. G. Proakis. *Digital Communication*. McGraw-Hill, 3 edition, 1995.

[36] Y. Sato. A method of self-recovering equilization for multilevel amplitude-modulation systems. *IEEE Transactions on Communications*, COM-23:679–682, 1975.

[37] E. Serpedin and G. S. Giannakis. Blind channel identification and equalization with modulation-induced cyclostationarity. *IEEE Transactions on Signal Processing,*, SP-46:1930–1944, 1998.

[38] O. Shalvi and E. Weinstein. New criteria for blind deconvolution of non-minimum phase systems (channels). *IEEE Transactions on Information Theory*, IT-36,:312–321, 1990.

[39] O. Shalvi and E. Weinstein. Super-exponential methods for blind deconvolution. *IEEE Transactions on Information Theory*, IT-39:504–519, 1993.

[40] B. Sklar. *Digital Communications*. Prentice-Hall, Englewood Cliffs, New Jersey, 1988.

[41] D. T. M. Slock. Blind fractionally-spaced equalization, perfect-reconstruction filter banks and multichannel linear prediction. In *Proceedings of ICASSP'94*, pages 585–588, Adelaide, Australia, 1994.

[42] D. T. M. Slock and C. B. Papadias. Further results on blind identification and equalization in multiple FIR channels. In *Proceedings of ICASSP'95*, pages 1964–1967, Detroit, Michigan, 1995.

[43] L. Tong, G. Xu, and T. Kailath. Fast blind equalization via antenna arrays. In *Proceedings of ICASSP'93*, pages 272–275, 1993.

[44] L. Tong, G. Xu, and T. Kailath. Blind identification and equalization based on second-order statistics: A time domain approach. *IEEE Transactions on Information Theory*, IT-40:340–349, 1994.

[45] L. Tong, G. Xu, and T. Kailath. Blind identification and equalization based on second-order statistics: A frequency domain approach. *IEEE Transactions on Information Theory*, IT-41:329–334, 1995.

[46] J. R. Treichler and B. G. Agee. A new approach to multipath correction of constant modulus signals. *IEEE Transactions on Acoustics, Speech and Signal Processing*, ASSP-31:459–471, 1983.

[47] M. K. Tsatsanidis and G. B. Giannakis. Transmitter induced cyclostationarity for blind channel equalization. *IEEE Transactions on Signal Processing*, SP-45:1785–1794, 1997.

[48] J. K. Tugnait. On fractionally spaced blind adaptive equalization under symbol timing offsets using godard and related equalizers. *IEEE Transactions on Signal Processing*, SP-44:1817–1821, 1996.

[49] H. L. Van-Trees. *Detection, Estimation and Modulation Theory, Part I*. Wiley, New York, 1968.

[50] G. Xu, H. Liu, L. Tong, and T. Kailath. A least-squares approach to
     blind channel identification. *IEEE Transactions on Signal Processing*,
     SP-43:2982–2993, 1995.

3 BLIND SYSTEM IDENTIFICATION

J K Richardson and A K Nandi

Contents

3.1	Introduction		105
	3.1.1	Types of Linear Parametric Models	105
		AR Filters	105
		MA Filters	106
		ARMA Filters	106
	3.1.2	Modelling for System Identification	107
	3.1.3	Applications	108
3.2	MA Processes		108
	3.2.1	MA Filters and Higher Order Statistics	109
		$c(q,k)$ Method	111
		$c(k,k)$ Method	113
		$c(q,k+i)$ Method	113
		Forward and Backward ARMA Approximations	114
		Second- and Third-Order Cumulant Methods	114
		GM89	116
		T-equation	117
		NKSK95	117
		AVC93	118
		FV93	119
	3.2.2	Second-, Third- and Fourth-Order Method, MN96	119
	3.2.3	Simulation Details	127
	3.2.4	Order Selection	127
	3.2.5	Cumulant Estimation	128

3.2.6   Specific Requirements of Algorithm MN96   129

        Initialisation   129

        Perturbation   129

        Convergence   130

3.2.7   Discussion and Results   130

3.2.8   Conclusions   134

3.3   ARMA Processes   136

3.3.1   ARMA Filters and Higher Order statistics   137

3.3.2   Residual Time Series (RTS) Methods   138

3.3.3   Q-slice Method   140

3.3.4   Double $c(q, k)$ Method   141

3.3.5   Impulse Response Method   142

3.3.6   ARMA system identification Method   142

        Estimation of the MA(q) Parameters of the ARMA(p,q)
            System   142

        A Mixed-Order Method for Estimation of the AR(p) Pa-
            rameters of an ARMA(p,q) System   143

3.3.7   ARMA Model Order Selection   145

3.3.8   Simulation Details   146

        Filter Coefficients   146

        Cumulant Estimates   146

        Model Order   147

3.3.9   Method GM89   147

        Q-Slice Method   147

        Method DCQK   148

3.3.10 Discussion and Results   149

        AR Parameter Estimation   149

        Overall Performance   153

3.3.11 Conclusions   160

References   162

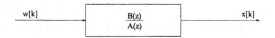

Figure 3.1: Model Transfer Function

3.1 Introduction

In this chapter a comparison of blind system identification methods for linear time-invariant (LTI) systems using HOS is presented [37]. These methods [35] use only the system output data to identify the system model under the assumption that the system is driven by an independent and identically distributed (i.i.d.) non-Gaussian sequence that is unobservable.

Blind (linear) system identification is the process which, with some general statistical assumptions on the input and the knowledge of the output, estimates the parameters of the (linear) system. Conventionally, second-order statistics, embracing correlation, covariance and power spectral analysis, were employed. Indeed, for a random process which exhibits a Gaussian distribution, hereafter known as a Gaussian random process, first- and second- moment analysis of the process characterises the distribution completely, capturing all the available information of the process. In some cases, the information of particular interest will be suppressed by the second-order statistics. For example, all second-order statistics suppress phase information such that any second-order statistical technique will in general identify a minimum phase model even if the real process is better described by a nonminimum phase model. This limitation is an intrinsic property of the statistics and can only be avoided by implementing higher-order statistical analysis. Below are listed three types of linear parametric models.

3.1.1 Types of Linear Parametric Models

There are three basic types of linear models: moving average (MA), autoregressive (AR), and autoregressive moving average (ARMA). These differ in the basic form of the model transfer function. If we consider a model with a rational transfer function we can describe it in the form of a numerator and a denominator polynomial as in figure (3.1).

AR Filters

An AR filter is recursive with an infinite impulse response (IIR) and has only a nontrivial denominator polynomial i.e. $B(z) = 1$ and $A(z) = 1 +$

$a(1)z^{-1} + a(2)z^{-2} + \cdots + a(p)z^{-p}$, where p is the order of the filter. As such all the zeros of an AR filter are located at the origin. The current output of an AR filter, $x[k]$, depends on the current input, $w[k]$, and previous outputs, $x[k-1], \ldots, x[k-p]$, giving the difference equation (3.1). As such it resembles a statistical regression and became known as an AutoRegressive filter by statisticians, a name which is used now across all fields.

$$x[k] + a(1)x[k-1] + \cdots + a(p)x[k-p] = w[k] \qquad (3.1)$$

This type of model is most frequently used because it is the simplest to design. If a stream of output data, or time series, is available a set of linear equations in the unknowns $a(1), \ldots, a(p)$ can be formed according to equation (3.1) and the parameters of the AR model can be estimated by solving the set of linear equations.

MA Filters

MA filters have a finite impulse response (FIR). The corresponding transfer function has only a nontrivial numerator polynomial leading to this type of filter being known interchangeably as an all-zero or moving average filter. Hence $A(z) = 1$ and $B(z) = b(0) + b(1)z^{-1} + b(2)z^{-2} + \cdots + b(q)z^{-q}$ where q is the order of the MA filter. Hence the output of an MA filter is a weighted average of the current and previous inputs described by difference equation (3.2). The parameter $b(0)$ is usually taken to be unity for system identification.

$$x[k] = b(0)w[k] + b(1)w[k-1] + \cdots + b(q)w[k-q] \qquad (3.2)$$

The simple transfer function of this type of filter conceals the fact that parameter estimation for an MA model is actually much more difficult than the estimation of the parameters of an AR model. Equation (3.2) can be rewritten in the form of the convolution sum of equation (3.3). Hence, solving for the MA parameters involves undoing the convolution sum which entails nonlinear algebraic manipulation.

$$x[k] = \sum_{i=0}^{q} b(i)w[k-i] \qquad (3.3)$$

ARMA Filters

The final type of filter possesses both nontrivial numerator and denominator polynomials. As such it is a combination of both an MA and an AR filter and is termed an ARMA filter. Hence $A(z) = 1 + a(1)z^{-1} + a(2)z^{-2} + \cdots + a(p)z^{-p}$ and $B(z) = 1 + b(1)z^{-1} + b(2)z^{-2} + \cdots + b(q)z^{-q}$ where p is the order of the denominator polynomial and is associated with the AR part of the filter and q is the order of the numerator polynomial associated with the MA part. Thus the difference equation for the model is given by equation (3.4) which can be rewritten in the form of equation (3.5).

$$x[k] + a(1)x[k-1] + \cdots + a(p)x[k-p] = w[k] + b(1)w[k-1] + \cdots$$
$$+ b(q)w[k-q] \quad (3.4)$$

$$\sum_{j=0}^{p} a(j)x[k-j] = \sum_{i=0}^{q} b(i)w[k-i] \quad (3.5)$$

This is the most flexible model and frequently requires far fewer parameters to accurately describe an output time series in terms of power spectral density, and for higher-order statistical methods in terms of phase response, than either an MA or AR model. However, its design is the most complicated of the three types of model.

3.1.2 Modelling for System Identification

System identification and power spectrum estimation have relied heavily on the use of a white noise driven, linear time-invariant (LTI) model of a random signal, termed the innovations representation. If the unobservable input, driving noise $w[n]$, is assumed to be Gaussian then employing second-order statistics to describe the output sequence will result in a minimum phase (MP) parametric model being identified. This minimum phase model may describe the magnitude response of the system accurately and still fail to describe the phase response. If the system was minimum phase then this approach will fully characterise the system. However, many real systems are not minimum phase and are either nonminimum phase (NMP) or maximum phase (MXP). The MP systems identified using second-order statistics will have an identical magnitude response but an incorrect phase response. This is because second-order statistics explicitly suppress phase information and are as such phase blind.

In order to characterise an NMP or MXP system completely higher-order statistics (HOS) are often employed, where the term HOS applies to all moments and cumulants of order 3 or higher. In addition, the driving noise, $w[n]$, can no longer be Gaussian since the higher-order cumulants and some moments of a Gaussian signal are identically zero, but it should be non-Gaussian.

3.1.3   Applications

In effect HOS can discriminate between comparative MP, NMP and MXP systems. In identifying a parametric model which describes the random output, the system transfer function is revealed and the system is identified. The system transfer function can then be used to enable prediction or control of the output by consideration of how the identified model would react under different inputs or by monitoring the output the input effectively becomes observable.

The use of HOS is not limited to parametric linear system identification. Rather parametric identification is the first step towards a further aim. A rich and varied field of applications and tools have been developed and include applications such as positional angle of source relative to observer for radar and sonar applications, and seismic applications. In addition, the use of HOS hold the promise of accurate nonlinear and non-Gaussian signal processing. The techniques and tools are still in their infancy compared to the wealth of expertise, knowledge and theory surrounding and supporting second-order statistics. As the tools emerge and become better understood and more robust, the list of applications, practical and theoretical (potential), is growing.

Examples of applications include speech coding and adaptive equalisation. In speech coding a best fit model is identified from the data, and the model is then used to create a predicted data sequence, the predicted sequence is then subtracted from the real sequence creating an error signal which is combined with the model coefficients to form the encoded speech signal which is relayed over the channel to be recovered at the receiver. In adaptive equalisation the constantly changing amplitude and delay distortions of a communications channel are compensated for automatically.

3.2   MA Processes

Various recursive and least squares methods for the identification of MA systems have been proposed using a variety of second, third and fourth order statistics and different 1-D cumulant slices [9], [11], [10], [48], [31] and [47]. [2] developed a new method which exploited all samples of the second and third order cumulants to reconstruct the unknown system impulse response. [28] and [49] proposed methods that depend on third order cumulants alone. Method [23] is remarkably different as it uses a combination of seconf, third and fourth order statistics. This aim in using this combination of cumulant orders is to support intrinsically and automatically all non-Gaussian input distributions. Hence method [23] can be used more widely and with more confidence regardless of prior knowledge of the system statistics. For an MA process the system output is related to the input by the convolution sum of the input with the rational system transfer function, $B(z)$. Two techniques are employed in the estimation of the filter coefficients: nonlinear methods and linear simplifications. Nonlinear solutions can be computationally expensive and may converge to a local minimum [2]. However, when these nonlinear methods are properly initialised the estimates obtained are generally better than the estimates obtained using linear methods. MN96 is a nonlinear method which currently uses the $c(q,k)$ formula [10] to initialise the algorithm. The method is remarkably insensitive to initialisation, however. Perturbation of the parameter estimates ensures that global, rather than local, minima are found. All $\{b(k)\}_{k=0}^{q}$ parameters are estimated simultaneously. Iteration of the method, using the previous parameter values to initialise the algorithm, is continued until the parameters settle to a stable value.

These methods assume prior knowledge of the MA model order, q. In fact, estimation of q from the time series is a substantial part of the system identification problem. Hence, model order selection has become an area of research in its own right.

3.2.1 MA Filters and Higher Order Statistics

There are many methods of blind system identification of MA processes using HOS. Some of these methods employ only third-order cumulants to identify the q MA parameters $\{b(k)\}_{k=0}^{q}$ whilst other methods use a combination of orders. In the following are summarised methods using third-order cumulants only described by [10], [17], [28] and [32], and the mixed-order methods of [11], [47], [26], [2] and [9]. The strengths and weaknesses of the different approaches according to current opinion are also highlighted. [25] gives a thorough review of some of the older existing parametric methods based on HOS and is an excellent introduction to the use of HOS.

Consider the system depicted schematically in figure (3.2). The noise-free signal, $x[k]$, is related to the driving noise, $w[k]$, by parameters $b(i)$ where i takes the values $0, 1, \ldots, q$ and q is the model order (i.e. the observed noise-free output is a weighted sum of the input only). The observed output, $y[k]$, is corrupted by additive Gaussian noise, $v[k]$. Note that it is usual for system identification methods to impose $b(0) = 1$. This introduces a scaling factor to all the parameters and does not affect system identification performance. Figure (3.3) shows a more generic representation of the MA(q) process where the filtering operation is depicted by the function $B(z)$. For an

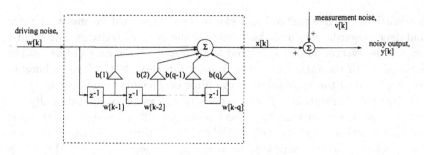

Figure 3.2: MA Modelling

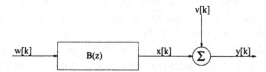

Figure 3.3: Notation for Cumulant Identification Methods

MA process $B(z)$ is a rational polynomial with order q equal to the number of previous inputs that affect the current output which gives the order of the process. The noise-free output, $x[k]$, is related to the input, $w[k]$, by constant weights, $\{b(i)\}_{i=0}^{q}$, given in equation (3.6). The corruption of this output by noise results in equation (3.7) where the observed noise-corrupted signal, $y[k]$, is summation of the noise-free signal and the additive noise, $v[k]$.

$$x[k] = \sum_{i=0}^{q} b(i)w[k - i] \qquad (3.6)$$

$$y[k] = x[k] + v[k] \qquad (3.7)$$

If higher order statistics are to be used to formulate a general relationship for the identification of the model parameters from the output of the system only, then the following conditions 1 and 2 must hold:

1. The driving noise, $w[k]$, is zero-mean, independent and identically distributed (i.i.d.) and non-Gaussian with $E\{w^2[k]\} = \gamma_{2w}^2$, $E\{w^3[k]\} = \gamma_{3w} \neq 0$ and $E\{w^4[k]\} - 3\gamma_{2w} = \gamma_{4w} \neq 0$.

2. The measurement noise, $v[k]$, is assumed to be zero-mean, i.i.d., and independent of $w[k]$. In addition it is assumed to be Gaussian in distribution with $E\{v^2[k]\} = \sigma_v^2$, $E\{v^3[k]\} = \gamma_{3v} = 0$ and $E\{v^4[k]\} - 3\sigma_v^2 = \gamma_{4v} = 0$.

Let $c_{3x}(m, n)$ represent the third order cumulant of the observed noise-free signal at lags m and n. Equation (3.8) relates the third-order cumulants at the specified combination of lags to the MA parameters $\{b(i)\}_{i=0}^{q}$ and the skewness of the input to the system, γ_{3w}.

$$c_{3x}(m, n) = \gamma_{3w} \sum_{k=0}^{q} b(k)b(k+m)b(k+n) \tag{3.8}$$

This relation is the basis for all methods of blind system identification which make use of third-order statistics. The corresponding relations for the covariance sequence and fourth-order cumulants follow intuitively and are given in equations (3.9) and (3.10) respectively.

$$c_{2x}(m) = \gamma_{2w} \sum_{k=0}^{q} b(k)b(k+m) \tag{3.9}$$

$$c_{4x}(m, n, s) = \gamma_{4w} \sum_{k=0}^{q} b(k)b(k+m)b(k+n)b(k+s) \tag{3.10}$$

$c(q, k)$ Method

[10] first formulated the c(q,k) approach which provides a closed form solution to the Moving Average (MA) parameter estimation problem.

Consider the system of equation (3.7) where $x[k]$ is given by equation (3.6) and $v[k]$ is Gaussian noise which is independent of $w[k]$. Assuming that $w[k]$ is third-order white, that is to say that the skewness of the system is finite, the third-order cumulant of $y[k]$ can be written in the form of equation (3.11) where the upper limit has been reduced to reflect the fact that the summation terms will be zero for $b(i)$ where $i > q$ since samples of $y[k]$ more than q lags apart are uncorrelated. Whilst $m \geqslant n$ the upper limit can therefore be changed from q to $(q - m)$ without loss of information.

$$c_{3y}(m, n) = \gamma_{3w} \sum_{i=0}^{q-m} b(i)b(i+m)b(i+n) \qquad m \geqslant n \geqslant 0 \tag{3.11}$$

Evaluating equation (3.11) for $m = q, n = k$ results in equation (3.12) which can be rearranged to give the relation for $b(k)$ in terms of $c_{3y}(q, k), \gamma_{3w}, b(0)$ and $b(q)$ of equation (3.13).

$$c_{3y}(q, k) = \gamma_{3w} b(0)b(q)b(k) \qquad 0 \leqslant k \leqslant q \tag{3.12}$$

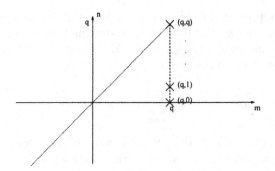

Figure 3.4: Cumulants Used in the c(q,k) Formula

$$b(k) = \frac{c_{3y}(q,k)}{\gamma_{3w}b(0)b(q)} \tag{3.13}$$

Setting $m = q$, $n = 0$ in equation (3.11) results in equation (3.14).

$$c_{3y}(q,0) = \gamma_{3w}b^2(0)b(q) \tag{3.14}$$

Equation (3.15) relating the impulse response of the MA model to the third-order cumulants of the system output is formed by combining the expressions for $b(k)$ and $c_{3y}(q,0)$ and assuming $\gamma_{3w} \neq 0$. Recalling $b(0) = 1$ the unknown filter coefficients may be calculated from the estimated third-order cumulants. The c(q,k) formula uses the vertical cumulant slice shown in figure (3.4) and requires exact knowledge of the MA order. Use of an incorrect model order causes the identification method to deteriorate rapidly especially in the case of overestimation. Theoretically the cumulants at lags greater than the system order will be zero. In practice, the estimated cumulants will be very small due to finite data lengths and the division of such numbers results in a completely erroneous coefficient estimate. As such, the importance of correctly selecting the model order cannot be overemphasised. The selection of an MA model order is the subject of research in its own right. In addition the method uses only two cumulant lags for the estimation of each b(k), the minimum possible information. This lack of redundancy means that the effect of additive noise on the cumulant estimates is in no way smoothed.

$$\frac{b(i)}{b(0)} = \frac{c_{3y}(q,i)}{\gamma_{3w}b^2(0)b(q)} = \frac{c_{3y}(q,i)}{c_{3y}(q,0)} \qquad i = 0,1,\ldots,q \tag{3.15}$$

This simple method of parameter estimation is used to initialise the novel mixed-order method, MN96.

$c(k,k)$ **Method**

The c(q,k) method can be modified to yield an equation using the diagonal slice of the third order cumulants. This method is obtained by setting $n = m$ in (3.11) generating a relation between the diagonal cumulant slice $c_3(k,k)$ and the MA parameters to be estimated.

$$c_{3y}(m,m) = \gamma_{3w} \sum_{i=0}^{q} b(i)b^2(i+m) \qquad m = -q,\ldots,0,\ldots,q \qquad (3.16)$$

[17] suggested two methods for the estimation of the MA coefficients based on the c(k,k) formula: a nonlinear least-squares approach, and a linear programming approach. The nonlinear approach minimises the performance criterion of equation (3.17) with respect to the $(q + 1)$ unknowns $\{b(i)\}_{i=1}^{q}$ and γ_{3w}.

$$\sum_{m=-q}^{q} \left[c_{3y}(m,m) - \gamma_{3w} \sum_{i=0}^{q} b(i)b^2(i+m) \right]^2 \qquad (3.17)$$

The linear programming approach uses an autocovariance based method to estimate the spectrally equivalent minimum phase (SEMP) system using only magnitude information. The roots of the SEMP system are then reflected inside or outside the unit circle until all combinations are exhausted. Equation (3.17) is calculated for each combination of $b(k)$ re-introducing the phase information contained in the third-order statistics and the set of roots that result in the minimum performance criterion value are chosen as the optimal MA parameters. However, any SEMP method requires an exhaustive search of all possible combinations of zero locations. The computational overhead is therefore very high and rapidly increases with model order. The use of a SEMP method is therefore compromised by the exhaustive search procedure and is not considered further.

$c(q,k+i)$ **Method**

[28] proposed three techniques based on a c(q,k+i) equation for blind system identification. The method replaces k with $(k+i)$ in the c(q,k) equation (3.12) giving equation (3.18).

$$c_{3y}(q,k+i) = \gamma_{3w}b(0)b(q)b(k+i) \qquad 0 \le (k+i) \le q \qquad (3.18)$$

Making suitable substitutions and evaluating over $0 \le i \le q$ the relation expressing $c_{3y}(k,k)$ in terms of $c_{3y}(q,k+i)$ of equation (3.19) is obtained which forms the basis of the algorithms described.

$$\sum_{i=1}^{q} b(i)c_{3y}^{2}(q,k+i) + b(0)c_{3y}^{2}(q,k) = \gamma_{3w}b^{2}(0)b^{2}(q)c_{3y}^{(}k,k) = -c_{3y}^{2}(q,k)$$

$$\text{for} \quad |k| \leq q$$

$$(3.19)$$

Forward and Backward ARMA Approximations

[32] considered splitting the numerator polynomial of an NMP MA system into two parts such that the first includes all zeros inside the unit circle and reflects the MP aspects of the system and the second includes all zeros outside the unit circle such that the transfer function $B(z)$ is partitioned, $B(z) = I(z^{-1})O(z)$. The resulting model may be approximated by a causal (forward) ARMA model or an anticausal (backward) ARMA model. Extension of the parameter estimation method based on ARMA approximations using fourth-order cumulants was proposed by [36] whilst [34] suggested the use of a noncausal AR approximation to the MA system identification problem.

Second- and Third-Order Cumulant Methods

Consider again the MA process depicted in figure (3.2) where

$$y[k] = x[k] + v[k]$$

and

$$x[k] = \sum_{i=0}^{q} b(i)w[k-i]$$

and $w[k]$ is an i.i.d. (independent identically distributed), zero-mean, non-Gaussian process and the noise $v[k]$ is zero-mean, i.i.d. process independent of $w[k]$. Evaluating the third-order cumulant sequence of $x[k]$ at lags m and $n = (m+\rho)$ and the auto-covariance sequence at lag m give equations (3.20) and (3.21).

$$c_{3x}(m,n) = \gamma_{3w}\sum_{k=0}^{q} b(k)b(k+m)b(k+m+\rho) \qquad (3.20)$$

$$c_{2x}(m) = \gamma_{2w}\sum_{k=0}^{q} b(k).b(k+m) \qquad (3.21)$$

In order to obtain a relation between the filter coefficients and second- and third- order cumulants a transform to the z-domain is made and the common factor, $B(z^{-1})$, is eliminated from the two transformed equations creating a single relation between the autocovariance sequence the third-order cumulants and the unknown filter coefficients. The z-transform of the second-order relation is straightforward and gives equation (3.22).

$$C_{2x}(z) = \gamma_{2w} B(z) B(z^{-1}) \qquad (3.22)$$

However, in order to generate a common factor the third-order relation is first partitioned in the form of equation (3.23) where $g(k; \rho) = b(k)b(k + \rho)$ and ρ is a constant.

$$c_{3x}(m, n) = \gamma_{3w} \sum_{k=0}^{q} b(k)g(k + m; \rho) \qquad (3.23)$$

Letting $G(z; \rho)$ denote the z-transform of the sequence $\{g(k; \rho)\}$ and $C_{3x}(z; \rho)$ represent the z-transform of the third order cumulant sequence $\{c_{3x}(m, m + \rho)\}$ the z-domain transform of equation (3.23) is equation (3.24)

$$C_{3x}(z; \rho) = \gamma_{3w} B(z^{-1}) G(z; \rho) \qquad (3.24)$$

where

$$G(z; \rho) = B(z) * [z^{\rho}.B(z)] = \sum_{k=0}^{q} g(k; \rho).z^{-k}$$

and

$$B(z) = \sum_{k=0}^{q} b(k) z^{-k}$$

and * denotes the convolution operation. Eliminating $B(z^{-1})$ from equations (3.24) and (3.22) gives equation (3.25)

$$G(z; \rho) C_{2x}(z) = \epsilon_3 B(z) C_{3x}(z; \rho) \qquad (3.25)$$

where

$$\epsilon_3 = \frac{\gamma_{2w}}{\gamma_{3w}}$$

Equation (3.26) is obtained by reverting to the time-domain by taking the inverse z-transform.

$$\sum_{k=0}^{q} b(k)b(k + \rho)c_{2x}(m - k) = \epsilon_3 \sum_{k=0}^{q} b(k).c_{3x}(m - k, m - k + \rho) \qquad (3.26)$$

A similar approach using second and fourth order statistics, [47], results in equation (3.27) where $m \geqslant n \geqslant s$ and $n = m + \rho, s = m + \theta$

$$\sum_{k=0}^{q} b(k)b(k + \rho)b(k + \theta)c_{2x}(m - k) =$$

$$= \epsilon_4 \sum_{k=0}^{q} b(k)c_{4x}(m - k, m - k + \rho, m - k + \theta) \quad (3.27)$$

and

$$\epsilon_4 = \frac{\gamma_{2w}}{\gamma_{4w}}$$

In order to solve the parameter estimation problem, the cumulants of the noise-corrupted, observed output, $y[k]$, not the noise-free $x[k]$ must be related to the parameters. Where relevant this issue is addressed alongside the details of the system identification technique.

GM89

[11] used (3.26) with $\rho = 0$ resulting in a method which combimes a diagonal slice of the third-order cumulant sequence of $x[k]$ and the autocovariance of $x[k]$ leading to (3.28) which is solved by least-squares minimisation using $m = -q, \ldots, 0, \ldots, 2q$.

$$c_{2x}(m) + \sum_{i=1}^{q} b^2(i)c_{2x}(m - i) =$$

$$= \epsilon_3 \left[c_{3x}(m, m) + \sum_{i=1}^{q} b(i)c_{3x}(m - i, m - i) \right] \quad (3.28)$$

with $\epsilon_3 = \gamma_{2w}/\gamma_{3w}$ and $b(0) = 1$.

[11] suggested both a recursive solution and a linear least squares solution to equation (3.28). However, both these methods were found to fail under certain circumstances. [47] suggested a revision to the method which effectively overcame the problems. The method of [11] with the fix suggested by [47] is commonly used to benchmark results and will be used throughout the results section. The combined method of [11] and [47] will be referred to as the GMT method. [41] generalised the GMT method to use cumulants of two, arbitrary orders. Both single slice and multiple slice methods are discussed.

[11] also proposed separate methods for selecting the model order of an MA system and blind system identification of an ARMA system. The reader should be aware that these methods are also used in comparing with available techniques. The reference to [11] may therefore be in the context of a reference to MA blind system identification, MA model order determination or ARMA system identification. The attention of the reader is therefore drawn to this issue.

T-equation

[47] also suggested an alternative approach to the solution of equation (3.26) by setting $\rho = q$, as opposed to the [11] approach where $\rho = 0$, equation (3.29) is formed which can be solved recursively by setting $m = -q, -q+1, \ldots, -1$ where $\epsilon'_3 = b(q)/\epsilon_3$. Furthermore the effect of additive noise on the system output was considered and equation (3.29) includes the effect of such additive noise.

$$\sum_{k=0}^{q} b(k)c_{3y}(k-m,q) - \epsilon'_3[c_{2y}(m) - \gamma_{2v}\delta(m)] = -c_{3y}(-m,q) \qquad (3.29)$$

[47] also extended the relation to give the second- and fourth-order solution of equation (3.30).

$$\sum_{k=0}^{q} b(k)b(k+\rho)b(k+\theta)c_{2x}(m-k) = \epsilon_4 \sum_{k=0}^{q} b(k)c_{4x}(k-m,\rho,\theta) \qquad (3.30)$$

where $\epsilon_4 = \gamma_{2w}/\gamma_{4w}$. Expressing (3.30) in terms of the observed signal, $y[k]$ gives (3.31)

$$\sum_{k=0}^{q} b(k)c_{4y}(k-\tau,q,q) - \epsilon'_4\left[c_{2y}(\tau) - \gamma_{2v}\delta(\tau)\right] = -c_{4y}(-\tau,q,q) \qquad (3.31)$$

where $\epsilon'_4 = b^2(q)/\epsilon_4 = \gamma_{4w}b^2(q)/\gamma_{2w}$.

NKSK95

In order to perform system identification in the presence of noise the number of equations that can be formed using the GMT method is reduced [47], since terms affected by the noise must be eliminated. The result of eliminating such terms is a smaller range of suitable m, thus the number of equations in the least squares solution is reduced. [26] suggested a third- and fourth-order method to the system identification problem. By eliminating second-order statistics the Gaussian noise will be automatically removed by the use of higher-order cumulants. The disadvantage in employing the higher-order cumulants is the requirement for longer data lengths and therefore an increase in processing time. However, results from [26] for coloured and white additive Gaussian noise shows the NKSK95 algorithm to outperform both the c(q,k) and GMT methods.

Equations (3.32) and (3.33) are used simultaneously giving $5q + 2$ equations in $2q + 2$ unknowns which are solved using least squares minimisation.

$$\sum_{i=1}^{q} b^3(i)c_{3y}(m-i,m-i) + \sum_{i=0}^{q} \epsilon b^2(i)c_{4y}(m-i,m-i,m-i) =$$

$$= -c_{3y}(m,m) \qquad -q \leqslant m \leqslant 2q \quad (3.32)$$

$$\sum_{i=1}^{q} b^3(i)c_{3y}(i-m,q) + \epsilon b(q)c_{4y}(m,m,m) =$$

$$= -c_{3y}(-m,q) \qquad -q \leqslant m \leqslant q \quad (3.33)$$

where $\epsilon = -\gamma_{3w}/\gamma_{4w}$.

AVC93

[2] suggested a second- and third-order method of blind system identification for FIR systems. The method described by [2] exploits a much larger data set of the output statistics than the GMT method (the method described by [11] with the fix described by [47]). The first method reconstructs the system impulse response, $h(n)$, from all the second- and third-order cumulants using the cumulant lags $-q \leqslant m, n \leqslant 2q$ and $m \geqslant n$ to create an overdetermined set of equations which is solved by singular value decomposition and least squares minimisation.

$$\sum_{i=0}^{q} b(i)c_3(m-i,n-i) = \sum_{k=0}^{q} \epsilon b(i)b(n-m+k)c_2(m-i) \qquad (3.34)$$

where

$$\epsilon = \gamma_{3w}/\gamma_{2w}$$

In effect, equation (3.34) is a generalised version of both the GMT and T-equations. The GMT equation can be recovered by considering the diagonal third-order cumulant slice, $c_3(m, m)$, and setting $m = n = \tau$ in equation (3.34) and multiplying both sides of the equation by $\frac{1}{\epsilon}$. The method described by [47] uses the 1-D slice $c_3(q, \tau)$ of third-order cumulants and is recovered by setting $m = q$ and $n = (m + q)$. The second method uses the power spectrum in conjuction with a 1-D slice of the bispectrum to perform system identification. This, and other, higher-order spectral methods are not considered in the scope of this book.

FV93

[9] reported a weighted subspace linear method for FIR blind system identification which unlike previous methods does not require prior estimation of the filter order. It relies on the use of singular value decompostion (SVD) to produce a well-conditioned solution to an underdetermined set of equations. The solution of any set of underdetermined equations is not unique but by further minimising the norm of the weight vector used in the formulation of the equations a solution can be found which uniquely identifies the system. This w-slice method estimates the weights in a minimum-norm sense such that the causal w-slice of cumulants at lag zero, i.e. $C_{rx}(0) = 0$ sum to unity. The filter coefficients are then estimated from this w-slice and minimum-norm weight vector. The proposed algorithm can be based on different sets of cumulant slices and is reported to give competitive results in terms of bias and variance. In addition overestimation of the order has little apparent effect on the performance of the algorithm in identifying the filter coefficients.

3.2.2   Second-, Third- and Fourth-Order Method, MN96

The method described here will be referred to as method MN96 [23]. Taking the diagonal cumulants $c_3(m, m)$ and $c_4(m, m, m)$ in equations (3.26) and (3.27) respectively gives the basis for the new mixed order method.

$$\sum_{k=0}^{q} b^2(k).c_2(m - k) = \epsilon_3 \sum_{k=0}^{q} b(k).c_3(m - k, m - k) \qquad (3.35)$$

and

$$\sum_{k=0}^{q} b^3(k).c_2(m - k) = \epsilon_4 \sum_{k=0}^{q} b(k).c_4(m - k, m - k, m - k) \qquad (3.36)$$

Changing the limits of the $c_2(.)$ summation in equations (3.35) and (3.36) and setting $b(0) = 1$, which may be done without any loss of generality, gives

$$c_2(m) + \sum_{k=1}^{q} b^2(k).c_2(m-k) = \epsilon_3 \sum_{k=0}^{q} b(k).c_3(m-k, m-k) \qquad (3.37)$$

and

$$c_2(m) + \sum_{k=1}^{q} b^3(k).c_2(m-k) = \epsilon_4 \sum_{k=0}^{q} b(k).c_4(m-k, m-k, m-k) \quad (3.38)$$

Eliminating $c_2(m)$ from equations (3.37) and (3.38) an equation encompassing second, third and fourth order cumulants is formed.

$$\epsilon_3 \sum_{k=0}^{q} b(k).c_3(m-k, m-k) - \sum_{k=1}^{q} b^2(k).c_2(m-k) =$$

$$= \epsilon_4 \sum_{k=0}^{q} b(k).c_4(m-k, m-k, m-k) - \sum_{k=1}^{q} b^3(k).c_2(m-k) \quad (3.39)$$

Note that the resulting equation is cubic in the unknown $b(k)$ which must be solved to identify the filter coefficients $\{b(k)\}_{k=1}^{q}$ with the proviso $b(0) = 1$. Equation (3.39) forms the basis of the novel mixed order method. Equation (3.39) is linearised with respect to $b(k)$ using the $c(q, k)$ formula derived by [10] and restated in equation (3.40).

$$b(k) = \frac{c_3(q, k)}{c_3(q, 0)} \qquad (3.40)$$

Problems arising from this simple estimation technique include erroneous sign estimation for $b(k)$. In order to avoid any erroreous sign detection equation (3.39) is linearised in terms of $b^2(k)$ rather than $b(k)$. Defining $s(k)$

$$s(k) = b^2(k) = \left(\frac{c_3(q, k)}{c_3(q, 0)} \right)^2 \qquad k = 1, 2, \ldots, q \qquad (3.41)$$

equation (3.39) is linearised resulting in equation (3.42).

$$\epsilon_3 \sum_{k=0}^{q} b(k).c_3(m-k, m-k) - \sum_{k=1}^{q} s(k).c_2(m-k) =$$

$$= \epsilon_4 \sum_{k=0}^{q} b(k).c_4(m-k, m-k, m-k) - \sum_{k=1}^{q} s(k).b(k).c_2(m-k) \quad (3.42)$$

Gathering the $b(k)$ enteries on the left and changing the limits of the $c_3(.)$ and $c_4(.)$ entries produces equation (3.43) relating second-, third- and fourth-order cumulants to the unknown filter coefficients.

$$\sum_{k=1}^{q} b(k).s(k).c_2(m-k) + \epsilon_3 \sum_{k=1}^{q} b(k).c_3(m-k, m-k)$$

$$- \epsilon_4 \sum_{k=1}^{q} b(k).c_4(m-k, m-k, m-k) =$$

$$= \epsilon_4.c_4(m, m, m) - \epsilon_3.c_3(m, m) + \sum_{k=1}^{q} s(k).c_2(m-k) \quad (3.43)$$

Note that the assumption that the output signal $x[k]$ was noise-free allowed the subscript x to be dropped from the cumulant notation to give $c_2(.)$, $c_3(.)$ and $c_4(.)$. In a noisy environment the measured output signal, $y[k]$, is a noise-corrupted version of $x[k]$ and the estimated cumulants are in fact $c_{2y}(.)$, $c_{3y}(.)$ and $c_{4y}(.)$. However, as long as all lags equal to zero i.e. $c_{2y}(0)$, $c_{3y}(0,0)$ and $c_{4y}(0,0,0)$ are excluded, and the additive noise is Gaussian in distribution then estimated cumulants from the observed time series will be identical to estimated cumulants from the noise-free time series. Thus

$$c_2(m) = c_{2x}(m) = c_{2y}(m), \qquad \text{for} \quad m \neq 0$$

$$c_3(m, m) = c_{3x}(m, m) = c_{3y}(m, m), \qquad \text{for} \quad m \neq 0$$

and

$$c_4(m, m, m) = c_{4x}(m, m, m) = c_{4y}(m, m, m), \qquad \text{for} \quad m \neq 0$$

The abbreviated notation $c_2(.)$, $c_3(.)$ and $c_4(.)$ can therefore continue to be used with the proviso $m \neq 0$ such that equation (3.43) is calculated over the range m

$$m = [2q \quad 2q-1 \quad \ldots \quad q+1 \quad -1 \quad -2 \quad \ldots \quad -q]^T.$$

However, setting $m = -q$ is used to establish a relation between ϵ_3 and ϵ_4 leaving a set of $2q-1$ overdetermined equations with q unknowns, $b(1)$, $b(2)$, ..., $b(q)$ which can be solved by least squares minimisation.

Setting $m = -q$ in equation (3.43) gives

$$\epsilon_4 = \epsilon_3 \left(\frac{c_3(-q, -q)}{c_4(-q, -q, -q)} \right) \quad (3.44)$$

Following [11] and setting $\tau = -q$ in equation (3.35) gives

$$b^2(0).c_2(-q) = \epsilon_3 \sum_{k=0}^{q} b(k).c_3(\tau - k, \tau - k)$$

but $b(0) = 1$ by definition and $c_2(-q) = c_2(q)$ giving equation (3.45)

$$\epsilon_3 = \frac{c_2(q)}{c_3(-q, -q)} \tag{3.45}$$

Hence ϵ_4 is given by equation (3.46).

$$\epsilon_4 = \frac{c_2(q)}{c_4(-q, -q, -q)} \tag{3.46}$$

The resulting overdetermined set of equations can be solved using least squares minimisation. Representing equation (3.43) in vector and matrix form $\mathbf{M}.x = r$ where bold font differentiates a matrix from a vector. The least squares solution is given by

$$\hat{x}_{ls} = (\mathbf{M}^T.\mathbf{M})^{-1}.\mathbf{M}^T.r \tag{3.47}$$

where

$$\mathbf{M} = \sum_{m} \sum_{k=1}^{q} \Big(s(k)c_2(m - k) + \epsilon_3 c_3(m - k, m - k)$$

$$- \epsilon_4 c_4(m - k, m - k, m - k) \Big) \tag{3.48}$$

$$x = [b(1) \quad b(2) \quad \ldots \quad b(q)]^T \tag{3.49}$$

and

$$r = \sum_{m} \Big[\epsilon_4 c_4(m, m, m) - \epsilon_3 c_3(m, m) + \sum_{k=1}^{q} s(k)c_2(m - k) \Big] \tag{3.50}$$

Since all cumulant slices are diagonal, e.g. $c_3(m, m)$, the third and fourth order cumulant notation will be abbreviated to $c_3(m)$ and $c_4(m)$ by which $c_3(m, m)$ and $c_4(m, m, m)$ are implied. Estimating $\mathbf{M}$ over the range $m = [2q \quad 2q - 1 \quad \ldots \quad q + 1 \quad -1 \quad -2 \quad \ldots \quad -q]^T$ gives the matrix of (3.51) which can be simplified by explicitly setting $c_3(m), c_4(m) = 0$ for $m > q$. Hence (3.51) becomes (3.52). Estimation of vector r over the same range of m results in (3.53) which is simplified to reflect the fact that theoretically cumulants further than q lags apart are identically zero resulting in (3.54).

$$
\mathbf{M} =
\begin{bmatrix}
\begin{array}{l} s(1).c_2(2q-1) \\ +\epsilon_3.c_3(2q-1) \\ -\epsilon_4.c_4(2q-1) \end{array} &
\begin{array}{l} s(2).c_2(2q-2) \\ +\epsilon_3.c_3(2q-2) \\ -\epsilon_4.c_4(2q-2) \end{array} &
\cdots &
\begin{array}{l} s(q).c_2(q) \\ +\epsilon_3.c_3(q) \\ -\epsilon_4.c_4(q) \end{array} \\[2em]
\begin{array}{l} s(1).c_2(2q-2) \\ +\epsilon_3.c_3(2q-2) \\ -\epsilon_4.c_4(2q-2) \end{array} &
\begin{array}{l} s(2).c_2(2q-3) \\ +\epsilon_3.c_3(2q-3) \\ -\epsilon_4.c_4(2q-3) \end{array} &
\cdots &
\begin{array}{l} s(q).c_2(q-1) \\ +\epsilon_3.c_3(q-1) \\ -\epsilon_4.c_4(q-1) \end{array} \\[2em]
\cdots & \cdots & \cdots & \cdots \\[1em]
\begin{array}{l} s(1).c_2(q) \\ +\epsilon_3.c_3(q) \\ -\epsilon_4.c_4(q) \end{array} &
\begin{array}{l} s(2).c_2(q-1) \\ +\epsilon_3.c_3(q-1) \\ -\epsilon_4.c_4(q-1) \end{array} &
\cdots &
\begin{array}{l} s(q).c_2(1) \\ +\epsilon_3.c_3(1) \\ -\epsilon_4.c_4(1) \end{array} \\[2em]
\begin{array}{l} s(1).c_2(-2) \\ +\epsilon_3.c_3(-2) \\ -\epsilon_4.c_4(-2) \end{array} &
\begin{array}{l} s(2).c_2(-3) \\ +\epsilon_3.c_3(-3) \\ -\epsilon_4.c_4(-3) \end{array} &
\cdots &
\begin{array}{l} s(q).c_2(-q-1) \\ +\epsilon_3.c_3(-q-1) \\ -\epsilon_4.c_4(-q-1) \end{array} \\[2em]
\begin{array}{l} s(1).c_2(-3) \\ +\epsilon_3.c_3(-3) \\ -\epsilon_4.c_4(-3) \end{array} &
\begin{array}{l} s(2).c_2(-4) \\ +\epsilon_3.c_3(-4) \\ -\epsilon_4.c_4(-4) \end{array} &
\cdots &
\begin{array}{l} s(q).c_2(-q-2) \\ +\epsilon_3.c_3(-q-2) \\ -\epsilon_4.c_4(-q-2) \end{array} \\[2em]
\cdots & \cdots & \cdots & \cdots \\[1em]
\begin{array}{l} s(1).c_2(-q) \\ +\epsilon_3.c_3(-q) \\ -\epsilon_4.c_4(-q) \end{array} &
\begin{array}{l} s(2).c_2(-q-1) \\ +\epsilon_3.c_3(-q-1) \\ -\epsilon_4.c_4(-q-1) \end{array} &
\cdots &
\begin{array}{l} s(q).c_2(-2q+1) \\ +\epsilon_3.c_3(-2q+1) \\ -\epsilon_4.c_4(-2q+1) \end{array}
\end{bmatrix}
\tag{3.51}
$$

$$
M = \begin{bmatrix}
0 & 0 & \cdots & 0 & \cdots & \substack{s(q).c_2(q)\\+c_3.c_3(q)\\+c_4.c_4(q)} \\[1.2em]
0 & 0 & \cdots & \substack{s(q-1).c_2(q)\\+c_3.c_3(q)\\-c_4.c_4(q)} & \cdots & \substack{s(q).c_2(q-1)\\+c_3.c_3(q-1)\\+c_4.c_4(q-1)} \\[1.2em]
\substack{s(1).c_2(q)\\+c_3.c_3(q)\\-c_4.c_4(q)} & \substack{s(2).c_2(q-1)\\+c_3.c_3(q-1)\\-c_4.c_4(q-1)} & \cdots & \substack{s(q-1).c_2(2)\\+c_3.c_3(2)\\-c_4.c_4(2)} & \cdots & \substack{s(q).c_2(1)\\+c_3.c_3(1)\\-c_4.c_4(1)} \\[1.2em]
\substack{s(1).c_2(-2)\\+c_3.c_3(-2)\\-c_4.c_4(-2)} & \substack{s(2).c_2(-3)\\+c_3.c_3(-3)\\-c_4.c_4(-3)} & \cdots & \substack{s(q-1).c_2(-q)\\+c_3.c_3(-q)\\-c_4.c_4(-q)} & \cdots & 0 \\[1.2em]
\substack{s(1).c_2(-3)\\+c_3.c_3(-3)\\-c_4.c_4(-3)} & \substack{s(2).c_2(-4)\\+c_3.c_3(-4)\\-c_4.c_4(-4)} & \cdots & 0 & \cdots & 0 \\[1.2em]
\vdots & \vdots & \vdots & \vdots & & \vdots \\[0.8em]
\substack{s(1).c_2(-q)\\+c_3.c_3(-q)\\-c_4.c_4(-q)} & 0 & \cdots & 0 & \cdots & 0
\end{bmatrix}
\tag{3.52}
$$

$$
r =
\begin{bmatrix}
s(1).c_2(2q-1) & +s(2).c_2(2q-2) & +\cdots+ & +s(q).c_2(q) & +\epsilon_4.c_4(2q) & -\epsilon_3.c_3(2q) \\
s(1).c_2(2q-2) & +s(2).c_2(2q-3) & +\cdots+ & +s(q).c_2(q-1) & +\epsilon_4.c_4(2q-1) & -\epsilon_3.c_3(2q-1) \\
\cdots & \cdots & \vdots & \cdots & \cdots & \cdots \\
s(1).c_2(q) & +s(2).c_2(q-1) & +\cdots+ & +s(q).c_2(1) & +\epsilon_4.c_4(q+1) & -\epsilon_3.c_3(q+1) \\
s(1).c_2(-2) & +s(2).c_2(-3) & +\cdots+ & +s(q).c_2(-q-1) & +\epsilon_4.c_4(-1) & -\epsilon_3.c_3(-1) \\
s(1).c_2(-3) & +s(2).c_2(-4) & +\cdots+ & +s(q).c_2(-q-2) & +\epsilon_4.c_4(-2) & -\epsilon_3.c_3(-2) \\
\cdots & \cdots & \vdots & \cdots & \cdots & \cdots \\
s(1).c_2(-q) & +s(2).c_2(-q-1) & +\cdots+ & +s(q).c_2(-2q+1) & +\epsilon_4.c_4(-q) & -\epsilon_3.c_3(-q)
\end{bmatrix}
\tag{3.53}
$$

$$
r = \left[
\begin{array}{llllll}
s(q).c_2(q) & \cdots & & \cdots & & \\
s(q-1).c_2(q) & +s(q).c_2(q-1) & \cdots & & +\epsilon_4.c_4(-1) & -\epsilon_3.c_3(-1) \\
s(q-1).c_2(2) & +s(q).c_2(1) & \cdots & & +\epsilon_4.c_4(-2) & -\epsilon_3.c_3(-2) \\
\vdots & \vdots & & & \cdots & \cdots \\
s(1).c_2(q) & +s(2).c_2(q-1) & +\cdots & \cdots & & \\
s(1).c_2(-2) & +s(2).c_2(-3) & +\cdots & +s(q-1).c_2(-q) & +\epsilon_4.c_4(-1) & -\epsilon_3.c_3(-1) \\
s(1).c_2(-3) & +s(2).c_2(-4) & +\cdots+ & & +\epsilon_4.c_4(-2) & -\epsilon_3.c_3(-2) \\
\vdots & \vdots & \vdots & & \cdots & \cdots \\
s(1).c_2(-q) & \cdots & \cdots & \cdots & +\epsilon_4.c_4(-q) & -\epsilon_3.c_3(-q)
\end{array}
\right] \tag{3.54}
$$

Table 3.1: Filter Coefficients for Systems Modelled

	b(0)	b(1)	b(2)	b(3)	Phase
b1	1	-2.05	1	-	NMP
b2	1	-2.333	0.667	-	NMP
b4	1	0.9	0.385	-0.771	MP
b7	1	-1.1314	0.64	-	MP
b10	1	-2.2	1.7325	0.49	NMP
bnksk	1	-1.40	0.98	-	MP
bfv93	1	-1.13	0.60	-	MP

3.2.3 Simulation Details

M Monte Carlo runs of K samples with the specified input distribution were generated using the rpiid package in the HOSA toolbox of MATLAB [44]. The necessary output data was then created by filtering the data given the set of filter coefficients to which Gaussian noise, either coloured or white, was added to give the required signal-to-noise ratio. This output data was used to perform blind system identification of the filter coefficients using a variety of MA parametric methods. For convenience all filters are assumed to have $b(0) = 1$.

The filter coefficients are defined in table (3.1) below where NMP corresponds to nonminimum phase filters and MP corresponds to minimum phase filters. Filters b1-b10 are taken from a variety of references and are frequently used in the literature to examine the performance of different algorithms. Filter *bnksk* was used in both [26] and [9]. However, different data lengths, K, and numbers of Monte Carlo runs were used. Filter *bfv93* was used in [9]. These filters will be used to compare the method MN96 with the results of [26] and [9]. Both these papers compare results of their simulations with results obtained by applying the Giannakis/Mendel algorithm to identical data. Therefore, in order to benchmark results and for completeness all results are compared to the Giannakis/Mendel algorithm with Tugnait's fix, termed the GMT method, available in the HOSA toolbox of MATLAB.

3.2.4 Order Selection

The problem of order selection for MA processes is the basis of research work in its own right. [9] proposed a method of system identification that estimated the system order by overfitting of the model within the algorithm.

The extra parameters of the system are then found to average to zero over
the M Monte Carlo runs. For the MA(5) system for which this is demonstrated the resulting standard deviations of the overfitted parameters are
comparable with that of the correct zeros, whilst the mean values are indeed
zero. This would suggest that individual simulations using the method did
not always result in the correct model order being estimated. Averaged over
the 256 Monte Carlo of the simulation the result for extraneous filter coefficients amounted to zero. Effectively the method has identified the correct
model order more frequently than it has overestimated it, but has not always created the combined solution of model order and system identification
simultaneously correctly. This is no different to any method which uses a
model order determination algorithm prior to system identification in that
incorrect model order estimation will affect the system identified. Averaged
over a number of simulations, the model order will predominately be correctly identified resulting in an accurate model with good filter coefficients.
Overestimation will result in a higher order model with more filter coefficients which may not be negligeable when considered separately, but which
when considered as one single realisation of M realisations will quickly result
in the extra filter coefficients being zero mean.

3.2.5   Cumulant Estimation

Cumulants involve expectations and cannot be computed in an exact manner from the real data. They must be approximated in a similar way to the
approximation of an autocovariance sequence [25]. All simulations were performed using the simplest approach to cumulant estimation described in [33]
which involves computing the mean of the finite length data at the required
lags. Several, more complex, techniques have been suggested to improve cumulant estimates and aim to either reduce the variance of the estimates for a
fixed length data sequence or compute similar variance estimates from shorter
data lengths. See [27], [21], [30] and [50] and the references therein. [42] exploits the theoretical result for MA processes that cumulants with lags greater
than the model order are zero and uses the estimated, and non-zero, values
of these lags to revise the cumulant estimates. Any of these methods could
be used prior to system identification and may result in improved filter coefficient values being obtained. However, for consistency only the simplest of
cumulant estimators was used.

3.2.6 Specific Requirements of Algorithm MN96

Initialisation

For each Monte Carlo run an estimate is made of all the required second, third and fourth order cumulants and of ϵ_3 and ϵ_4. The initial filter coefficients $b(k)$ are estimated from the $c(q,k)$ algorithm which provides an initial linearisation method for equation (3.43). The least squares minimisation of the overdetermined set of equations yields a new estimate of the filter coefficients. These are squared to give an $s(k)$ estimate which is used in the subsequent iteration of equation (3.43) to update $b(k)$. This process of updating $s(k)$ for the linearisation of equation (3.43) is continued for each of the N iterations. During the iterative calculation of the filter coefficients, instability where filter coefficients change radically between successive iterations is undesirable and must be avoided. This is accomplished by the implementation of a damping factor to limit this change between successive iterations of the algorithm. A damping factor of 0.4 was found to be optimal for all filters considered in terms of the compromise between bias and variance of the parameter estimates.

Perturbation

During initial verification of the algorithm incorrect filter coefficients were occasionally returned. This was attributed to the solution being formed from a locally stable set of filter coefficients which were the globally stable, correct set of filter coefficients. A perturbation technique was implemented to improve the local solution rejection and was found to eliminate it in over 90% of all simulations. Perturbation was implemented in the manner described below.

If, during iteration, the filter coefficients settled to a set of values which were unchanged between successive iterations to three decimal places a single-kick perturbation was applied. A perturbation was also applied if the filter coefficients failed to settle within a predetermined number of iterations during any one Monte Carlo run.

All perturbations took the form of a single, simultaneous forced increment or decrement to the current values of $b(k)$ and hence to the values of $s(k)$ used in the subsequent iteration. All parameters were altered by the same, fixed amount regardless of current value. Unity step size perturbations were used with alternating sign such that successive perturbations possessed opposite signs. The aim of the perturbation was to dislodge the filter coefficients from their current values to ascertain whether they would settle back to a similar state. Over three perturbations, the set of filter coefficients would then be

Table 3.2: Run Information for b1, b2 and b7 using method MN96

	actual values		estimated mean ± std	
filter	b(1)	b(2)	b(1)	b(2)
b1	-2.05	1	-2.074 ± 0.535	1.027 ± 0.273
b2	-2.333	0.667	-2.104 ± 0.387	0.627 ± 0.249
b7	-1.1314	0.64	-1.135 ± 0.091	0.544 ± 0.071

declared local or global. Local states were rejected in favour of the more robust, global states.

Convergence

For each M Monte Carlo run, N iterations of the algorithm MN96 were performed. The iterative process allowed the estimated parameters to stabilise to a set of values prior to perturbation. The convergence from the initial parameters estimated from the c(q,k) method was found to be optimised for a damping factor of 0.4. This provided the optimal compromise between the number of iterations necessary to observe stablisation of the parameter values and the convergence error, which effectively forms the estimate bias. Various methods could be employed to improve convergence. For example, a gradient descent method where previous sets of estimated coefficient values were used in conjunction to provide the new intialisation of the algorithm would reduce the number of iterations until convergence. The different methods of convergence are not included here. In principle, the same convergence method could be used on a number of the different MA BSI algorithms. [25] and [22] discuss convergence and different techniques that can be used.

3.2.7   Discussion and Results

The results using the novel second-, third- and fourth- order method, MN96, for one set of 30 Monte Carlo runs on data relating to filters b1, b2 and b7 are summarised in table (3.2) and detail the estimated mean and standard deviation of the filter coefficients. Where sample length and input distribution are not explicitly stated a sample length of $K = 1024$ and a negative exponential input distribution of mean zero and variance one were used. Parts (a) – (d) of figures (3.5) show the magnitude and phase characteristics for the mean estimated filters over the normalised frequency range and compare the results of the system identification algorithm with the actual filter per-

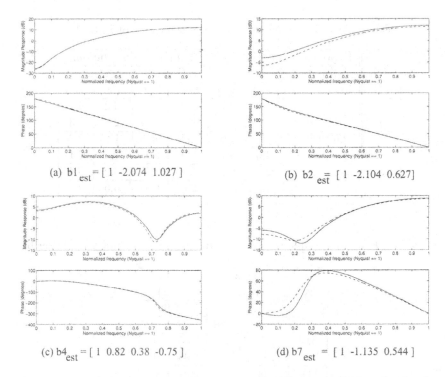

(a) b1$_{est}$ = [1 -2.074 1.027]

(b) b2$_{est}$ = [1 -2.104 0.627]

(c) b4$_{est}$ = [1 0.82 0.38 -0.75]

(d) b7$_{est}$ = [1 -1.135 0.544]

Figure 3.5: Frequency Domain Characteristics of Estimated Filters b1, b2, b4 and b7 (solid line actual system, broken line system estimated using MN96)

formance for filters b1, b2, b4 and b7 respectively. In each figure the solid line depicts the actual filter magnitude and phase characteristics whilst the broken line represents the filter identified by averaging the M Monte Carlo runs. Parts (a) and (d) of figure (3.5) are the result of simulations using a sample length of $K = 1024$ and a negative exponential input distribution whilst parts (b) and (c) of figure (3.5) are the result of simulations using an identical sample length but a Laplacian (symmetric) input distribution with mean zero and variance one. From figure (3.5) method MN96 is seen to correctly estimate the filter coefficients for both symmetric and asymmetric input distributions. Additive white Gaussian noise has little effect on the coefficient estimates for filter b1. Table (3.3) gives results from simulations for a data length, $K = 1024$ for both symmetrical and asymmetrical input distributions, namely a Laplacian distribution and a negative exponential distribution, over a range of signal-to-noise ratios (SNR) with results aver-

Table 3.3: Effect of Noise on Filter Coefficients of Filter b1

SNR	Exponential distribution				Laplacian distribution			
	estimated		% error		estimated		% error	
dB	b(1)	b(2)	b(1)	b(2)	b(1)	b(2)	b(1)	b(2)
∞	-2.07	1.03	1.0	2.7	-1.92	1.06	6.3	6.1
20	-2.11	0.99	3.0	1.3	-1.99	1.02	2.9	2.3
16	-2.02	1.03	1.6	3.0	-2.08	0.98	1.9	2.0
12	-2.14	1.07	4.6	6.7	-2.05	1.03	0	3.1
8	-2.00	1.06	2.4	5.6	-1.97	0.95	3.9	5.4
4	-2.08	0.90	1.5	10.2	-2.04	1.04	0.5	3.8

Table 3.4: Effect of Sample Length on Filter Coefficients and σ for Filter b1

K	mean values ± std	
	b(1)	b(2)
	-2.05	1.00
256	-1.91 ± 0.60	1.08 ± 0.60
512	-1.94 ± 0.37	1.05 ± 0.29
1024	-2.07 ± 0.53	1.03 ± 0.27
2048	-2.00 ± 0.29	1.07 ± 0.22
4096	-1.91 ± 0.20	1.00 ± 0.16
8192	-1.90 ± 0.12	1.00 ± 0.11

aged over 30 Monte Carlo runs. Other filters simulated exhibited the same level of noise rejection and similar performance in terms of mean error and standard deviation over a similar range of SNR. Performance is maintained at an approximately constant level until signal-to-noise levels drop to below 4dB when performance begins to deteriorate and the estimates of the filter coefficients are significantly affected. The effect of varying the sample length, K, on the mean estimates of the filter coefficients and on their standard deviation, σ, is shown in table (3.4) for filter b1. The results show that the mean parameter estimate varies only marginally with K, a result mimicked by other filters. However, all systems exhibit the decrease of standard deviation with increasing K predicted by statistical theory and fall away exponentially as shown for filter b1 in table (3.4).

Tables (3.5) and (3.6) show a comparison of filter coefficients estimated

Table 3.5: Filter b1, Method MN96 vs Giannakis/Mendel Method

signal/noise	MN96 mean ± std		GMT mean ± std	
dB	b(1)	b(2)	b(1)	b(2)
∞	-2.07 ± 0.54	1.03 ± 0.27	-1.92 ± 1.81	0.77 ± 0.76
20	-2.11 ± 0.46	1.06 ± 0.30	-2.23 ± 1.36	0.81 ± 1.06
16	-2.02 ± 0.38	1.02 ± 0.34	-2.41 ± 4.29	0.54 ± 1.73
12	-2.14 ± 0.60	1.06 ± 0.33	-5.04 ± 19.05	5.15 ± 25.86
8	-2.00 ± 0.57	1.06 ± 0.48	-1.59 ± 3.00	0.54 ± 1.04
4	-2.08 ± 0.53	0.90 ± 0.18	-2.62 ± 1.76	0.92 ± 0.87

Table 3.6: Filter b2, Method MN96 vs Giannakis/Mendel Method

signal/noise	MN96 mean ± std		GMT mean ± std	
dB	b(1)	b(2)	b(1)	b(2)
∞	-2.10 ± 0.39	0.63 ± 0.25	-3.93 ± 7.16	0.22 ± 1.55
20	-2.65 ± 0.83	0.74 ± 0.56	-0.86 ± 15.90	0.68 ± 2.93
16	-2.26 ± 0.86	0.67 ± 0.61	-4.41 ± 10.92	-0.03 ± 1.92
12	-2.23 ± 0.26	0.74 ± 0.50	-3.16 ± 5.82	0.26 ± 0.96
8	-2.23 ± 0.50	0.77 ± 0.57	-4.58 ± 11.40	0.21 ± 2.14
4	-2.78 ± 0.70	0.60 ± 0.49	-2.64 ± 11.11	0.08 ± 2.50

using MN96 and the GMT method. The GMT method is not a robust method and should rarely, if ever, be used in practice. However, this method is frequently used for benchmarking results and is included here in that specific context. Filters b1 and b2 show the typical improvement of the coefficients achieved using MN96. The improvement to the standard deviation, σ, of the M sets of filter coefficients produced is also shown. Figure (3.6) gives a comparison of the percentage error in the filter coefficients for filters b1, b2 and b7 for methods MN96 and GMT. Table (3.7) shows results for the system bnksk = [1 -1.40 0.98] with $K = 5120$ and 30 Monte Carlo runs. This filter was used for simulation by [26] and the results in columns GMT1 and NKSK95 are taken directly from [26]. No implementation of this algorithm was made. Assessment of performance is therefore made on statistically similar, rather than identical, output data sequences. Where coloured noise was used, this was generated by

$$v[k] = e[k] + 0.5e[k-1] - 0.25e[k-2]$$

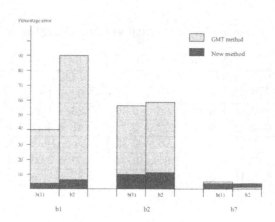

Figure 3.6: Comparison of Percentage Errors for Filters b1, b2 and b7

where e[k] is i.i.d. Gaussian. GMT1 of [26] is in fact the method referred to as the GMT algorithm in this book. Results for the method MN96 show an improvement in the mean estimated parameter value compared to the results of [26] for lower SNR with similar standard deviations for a range of signal-to-noise ratios. Both MN96 and NKSK95 show a marked improvement on the performance of the GMT method. For both these methods the improvement is most significant for additive coloured noise. [9] gives results for two MA(2) systems with $K = 100$ averaged over 1000 Monte Carlo runs and with an infinite SNR. Results obtained for method MN96 are given alongside the results taken directly from example 1 of [9] in table (3.8). For these two MA(2) systems method MN96 shows a performance equal to that of FV93. However, both these filters are MP and further simulation would be needed to separate the differences in performance which may be offered by the different methods.

3.2.8   Conclusions

This section has introduced a variety of parametric methods for blind system identification of MA processes based on higher-order statistics. The simulations using MN96 were undertaken for a variety of filter configurations which encompassed both MP and NMP systems. Results show that method MN96 works effectively for both minimum phase and nonminimum phase systems. It improves upon the results obtained for the GMT method in terms of improved coefficient estimation, improved stability of the estimates with reduced standard deviation. Further comparisons were made with re-

Table 3.7: Comparison of Results for b = [1 -1.40 0.98]

b(1) or b(2)	SNR	GMT1	NKSK95	MN96
b(1) = -1.40	∞	-1.402 ± 0.073	-1.465 ± 0.163	-1.372 ± 0.087
b(2) = 0.98		0.994 ± 0.050	0.980 ± 0.069	0.974 ± 0.095
b(1) = -1.40	10dB white	-1.305 ± 0.069	-1.430 ± 0.138	-1.361 ± 0.071
b(2) = 0.98		1.004 ± 0.053	0.956 ± 0.086	0.964 ± 0.091
b(1) = -1.40	10dB coloured	-1.306 ± 0.075	-1.422 ± 0.109	-1.393 ± 0.092
b(2) = 0.98		1.014 ± 0.053	0.951 ± 0.081	0.939 ± 0.089
b(1) = -1.40	5dB white	-1.255 ± 0.190	-1.331 ± 0.351	-1.416 ± 0.099
b(2) = 0.98		1.188 ± 0.202	0.926 ± 0.111	0.935 ± 0.105
b(1) = -1.40	5dB coloured	-1.292 ± 1.005	-1.327 ± 0.176	-1.392 ± 0.106
b(2) = 0.98		1.525 ± 1.130	0.913 ± 0.105	0.997 ± 0.161

Table 3.8: Comparison of Results for Two MA(2) Systems of [FV93]

b(1) or b(2)	GMT1	FV93	MN96
b(1) = -1.40	-1.46 ± 0.26	-1.41 ± 0.16	-1.42 ± 0.19
b(2) = 0.98	0.98 ± 0.08	0.99 ± 0.17	0.98 ± 0.14
b(1) = -1.13	-1.19 ± 0.26	-1.13 ± 0.09	-1.18 ± 0.07
b(2) = 0.60	0.59 ± 0.06	0.61 ± 0.07	0.59 ± 0.02

sults from [9] and [26]. Both filters bnksk and bavc are MP. Neither of these papers gives results for a NMP system. MN96 is shown to have improved performance over the method of [26] and similar performance to that of [9]. MN96 has been shown to have performance equal to that of [9] and better than [11] and [26] when an asymmetric input noise distribution is used. However, MN96 has also been seen to perform well for symmetric input distributions which differentiates it from the other methods. Results for method MN96 using a Laplacian input distribution and a negative exponential input distribution are given and both show the same high level of performance in terms of accuracy of the mean estimate and low variance.

In addition relatively short output data lengths, of the order of K=1204 for example, appear to produce good parameter estimates. Hence the method could be used for a non-stationary system by generating a series of system models which are constantly updated to give a best fit model at any one time. However, higher order systems require a longer output data length to perform as effectively as lower order systems. Higher order systems, therefore, require a greater degree of stationarity than lower order systems. [42] proposed a method for MA processes which exploits the difference between the estimated cumulants and those cumulants that should theoretically be zero to improve the cumulant estimates.

3.3   ARMA Processes

This section contains results of an ARMA(p,q) method of blind system identification which uses a residual time series method and the MA(q) method of blind system identification, MN96, proposed in an earlier section. A new method, JKR96 described in this section, of automatically selecting a combination of second-, third- and fourth-order cumulants for the estimation of the AR(p) parameters is described and the overall ARMA(p,q) process estimates both AR(p) and MA(q) parameters satisfactorily for both symmetric and asymmetric input distributions. The abillty to identify systems driven by inputs with both symmetrical and asymmetrical input distributions differentiates method JKR96 from other ARMA(p,q) methods. The only restriction placed on the driving noise is that it must possess a non-Gaussian input distribution. This is the minimum possible requirement for the identification of a nonminimum phase system.

Various methods for ARMA(p,q) identification using higher-order statistics have been proposed. [16] proposed a system impulse response method which used both the conventional power spectrum and the bispectrum. [18], [19] extended the work to use the trispectrum. [11] proposed a residual time-series method (RTS) where the AR(p) parameters are estimated first, and formation of a residual time series allows the MA(q) parameters to be estimated. An RTS approach means that any AR(p) and MA(q) methods can be paired to give a suitable ARMA(p,q) method. [12] proposed the double c(q,k) method where a moving average method is applied twice and the AR(p) and MA(q) parameters are reconstructed from the result. [43] proposed a method requiring q-slices of cumulants to determine the MA(q) parameters. [46] used cumulant matching in conjunction with an optimisation approach to estimate the ARMA(p,q) parameters. Various methods exploiting Spectrally Equivalent Minimum Phase (SEMP) filters have been proposed that determine a minimum phase system which accurately reflects the magnitude characteristics of the output with frequency. The zeros of the SEMP filter are then reflected outside the unit circle one by one to give a number of zero location combinations. The combination which minimises some function is then taken to be the correct NMP filter. [45] proposed a method that distinguishes the true model from the corresponding SEMP model using second- and fourth-order statistics. [20] build on the method of [45] eliminating the need to perform deconvolution by exploiting linear higher order statistics. More recently [4] proposed using the kurtosis of the system to localise the zeros of the true NMP filter from its SEMP model. However, this method does require deconvolution. [6] proposed an adaptive third-order cumulant based recursive least squares algorithm which is applied to the residual time series left when the influence of the AR(p) part of the ARMA(p,q) model is removed. Results from the ARMA(p,q) method proposed here, JKR96, are compared with those from the methods described by [11], [12] and [43] since these methods also explicitly search for a NMP system rather than a SEMP system followed by a NMP system that optimises some criterion. Comparison with any of the SEMP methods is not explicitly made.

3.3.1 ARMA Filters and Higher Order statistics

Consider the system of figure (3.7) where the observable output of the system, $y[k]$, is a noise-corrupted version of the true output, $x[k]$. The noise-free output, $x[k]$, is a linear combination of past values of the output, $x[k-1], \ldots, x[k-p]$, and current and past values of the input, $w[k], w[k-1], \ldots, w[k-q]$, where p and q are the AR and MA model orders respectively. The output process, $x[k]$, satisfies difference equation (3.55) and assuming the output of the filter is corrupted by additive Gaussian noise the observable output, $y[k]$, is the sum of the noise-free output and the independent noise source, $v[k]$,

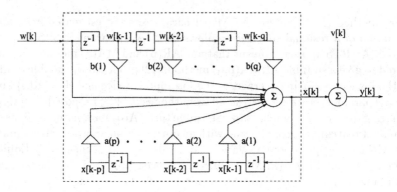

Figure 3.7: ARMA System Model

shown in equation (3.56).

$$\sum_{i=0}^{p} a(i)x[k-i] = \sum_{j=0}^{q} b(j)w[k-j] \qquad (3.55)$$

$$y[k] = x[k] + v[k] \qquad (3.56)$$

Methods which achieve estimation of the $\{a(i)\}_{i=0}^{p}$ and $\{b(i)\}_{i=0}^{q}$ parameters from the output of the system alone are termed blind system identification methods. The objective of blind system identification is to fit a parametric model to the output data of the system such that the model identifies the system. This objective requires the estimation of the ARMA parameters $\{a(k)\}_{k=0}^{p}$ and $\{b(k)\}_{k=0}^{q}$, and can be achieved using only the noise-corrupted output sequence, $y[k]$, if higher-order statistics, and specifically cumulants, are used.

3.3.2   Residual Time Series (RTS) Methods

The solution to the parameter estimation problem for ARMA system iden-tification can be generated in a two part process known as a Residual Time Series (RTS) method [13]. Consider the ARMA(p,q) process with general equation (3.55).

In this form equation (3.55) expresses the dependence of the output on current and previous inputs, and previous outputs. However, the output can also be expressed in terms of the impulse response in the frequency domain. Rewriting equation (3.55) in terms of a difference equation gives equation

(3.57).

$$a(0)x[k] + a(1)x[k-1] + \ldots a(p)x[k-p] = b(0)w[k] +$$
$$b(1)w[k-1] + \ldots b(q)w[k-q] \quad (3.57)$$

Note that in the system identification problem it is usual to set $a(0) = b(0) = 1$ which can be achieved without any loss of generality. Taking the z-transform of difference equation (3.57) with $a(0) = b(0) = 1$ results in equation (3.58).

$$H(z) = \frac{X(z)}{W(z)} = \frac{1 + b(1)z^{-1} + \cdots + b(q)z^{-q}}{1 + a(1)z^{-1} + \cdots + a(p)z^{-p}} \quad (3.58)$$

For $k = 0, 1, \ldots, q$ parameters $b(k)$ possesses a non-trivial solution. For $k > q$ the parameters $b(k) = 0$ such that the right hand side of equation (3.58) becomes zero. Expressing equation (3.58) in terms of these two conditions gives equation (3.59).

$$\sum_{i=0}^{p} a(i)h(k-i) = \begin{cases} b(k) & \text{if } 0 \geqslant k \leqslant q, \\ 0 & \text{if } k > q . \end{cases} \quad (3.59)$$

Hence assuming that the ARMA process does not contain all-pass factors and is free of pole-zero cancellations, the parameters $\{a(i)\}_{i=0}^{p}$ can be calculated by solving the overdetermined system of equations arising from setting $k > q$. The third-order cumulants of an ARMA system are given by equation (3.60). Substituting the third-order cumulants for the impulse response in equation (3.59) forms equation (3.61) which is used to solve for the AR parameters $\{a(i)\}_{i=1}^{p}$.

$$c_{3x}(m,n) = \gamma_{3w} \sum_{i=0}^{\infty} h(i)h(i+m)h(i+n) \quad (3.60)$$

$$\sum_{i=0}^{p} a(i)c_3(m-i,n) = 0 \qquad \text{for } m > q \quad (3.61)$$

A residual time series method is then applied where the influence of the AR parameters is extracted from the output signal leaving a time series that reflects only the $MA(q)$ process. This is implemented by applying the output $x[k]$ to the p-th order filter with transfer function $\hat{A}(z) = 1 + \sum_{i=1}^{p} \hat{a}(i)x[k-i]$ where $\hat{a}(i)$ is the estimate of the AR coefficients and $\hat{A}(z)$ is the resulting AR transfer function. The residual time series, $\hat{x}[k]$, is given by equation (3.62) and is an MA process of order q.

$$\hat{x}[k] = \sum_{i=0}^{q} b(i)w[k-i] \qquad (3.62)$$

Equation (3.62) can be solved by any MA method. However, [11] used the second- and third-order MA method described earlier as the GMT method and reference to method GM89 in the context of ARMA system identification will imply use of this specific method of MA identification.

3.3.3   Q-slice Method

[43] developed a method of ARMA system identification termed the q-slice method because q-slices of cumulant data, where q is the order of the MA part of the ARMA process, are required to identify the MA(q) parameters. The AR parameters are identified in an identical manner to the RTS method of [11] by solving equation (3.61). The impulse response of the system is then reconstructed using the estimated $a(i)$ parameters and the MA parameters can be estimated from the impulse response.

The impulse response of an ARMA process satisfies equation (3.63).

$$\sum_{i=0}^{p} a(i)h(k-i) = \sum_{j=0}^{q} b(j)\delta(k-j) = b(k) \qquad 0 \leqslant k \leqslant q \qquad (3.63)$$

For causal systems, the third-order cumulants of the output $x[k]$ are related to the system impulse response by equation (3.64).

$$c_{3x}(m,n) = \gamma_{3w} \sum_{i=0}^{\infty} h(i)h(i+m)h(i+n) \qquad (3.64)$$

Equations (3.63) and (3.64) can be combined to give equation (3.65) which forms the basis of MA parameter estimation problem.

$$\sum_{i=0}^{p} a(i)c_{3x}(m-i,k) = \gamma_{3w} \sum_{j=0}^{q} b(j)h(j-m+k)h(j-m) \qquad (3.65)$$

For a causal model, $b(i)$ is nonzero for $0 \leqslant i \leqslant q$ only. Evaluating equation (3.65) for $m = q$ and assuming $a(0) = b(0) = 1$ gives equation

(3.66). Substituting for $\gamma_{3w}b(q)$ in terms of third-order cumulants leads to equation (3.67) which is used to estimate the impulse response.

$$\sum_{i=0}^{p} a(i)c_{3x}(q-i,k) = \gamma_{3w}h(k)b(q) \qquad (3.66)$$

$$h(k) = \frac{\sum_{i=0}^{p} a(i)c_{3x}(q-i,k)}{\sum_{i=0}^{p} a(i)c_{3x}(q-i,0)} \qquad (3.67)$$

The impulse response and the AR parameters are then used to estimate the MA parameters from equation (3.68).

$$b(k) = \sum_{i=0}^{k} a(i)h(k-i) \qquad k = 1,\ldots,q \qquad (3.68)$$

For an MA process equation (3.67) reduces to the c(q,k) method of [11], which is a special case of the q-slice algorithm. Usually the AR parameters and the impulse response are computed in a single step by solving equation (3.69) for $k = 0,\ldots Q$ where $Q > q$.

$$\sum_{i=1}^{p} a(i)c_{3x}(q-i,k) - \gamma_{3w}b(q)h(k) = -c_{3x}(q,k) \qquad (3.69)$$

3.3.4 Double $c(q,k)$ Method

[12] describes a method of blind identification of an ARMA model from a discrete timeseries using HOS. The method is unusual in that it allows both causal and noncausal systems to be identified. The method is the least-squares solution to a quadratic model fitting of a sampled cumulant sequence [12]. Let $\alpha_3(i,j)$ and $\beta_3(i,j)$ be defined in terms of the AR(p) and MA(q) parameters of equations (3.70) and (3.71) respectively.

$$\alpha_3(i,j) = \sum_{k=0}^{p} a(k)a(k+i)a(k+j) \qquad (3.70)$$

$$\beta_3(i,j) = \sum_{k=0}^{q} b(k)b(k+i)b(k+j) \qquad (3.71)$$

Rewriting the general ARMA relation

$$\sum_{i=0}^{p} a(i)x[k-i] = \sum_{j=0}^{q} b(j)w[k-j]$$

in terms of $\alpha_3(i,j)$ and $\beta_3(i,j)$ gives equation (3.72) which employs both the redundant and nonredundant regions of support of $\beta_3(i,j)$ where $S(q) = \{0 \leqslant m \leqslant q, n \leqslant m\}$.

$$\sum_{i,j=-p}^{p} \alpha_3(i,j)c_3(m-i,n-j) = \begin{cases} 0 & \text{if } (m,n) \notin S(q), \\ \gamma_{3w}\beta_3(m,n) & \text{if } (m,n) \in S(q). \end{cases} \quad (3.72)$$

Equation (3.72) is solved for $\alpha_3(i,j)$ setting $m > q$ and subsequently solved for $\beta_3(i,j)$ using the estimated $\alpha_3(i,j)$.

3.3.5   Impulse Response Method

[16] proposed using the power spectrum to provide an estimate which includes magnitude information and the bispectrum to provide phase information such that NMP systems can be correctly identified. The power spectrum and input variance are estimated first using conventional second-order techniques described for example by [22]. The phase of the transfer function is reconstructed from a bispectrum estimate of the data using one of the nonparametric methods described by [5], [17], [3] or [24]. [16] gives the transfer function of the ARMA(p,q) as

$$H(w) = \left[\frac{C_{2x}(w)}{\gamma_{2w}}\right]^{\frac{1}{2}} exp\{j\phi(w)\}$$

where $C_{2x}(w)$ is the power spectrum of the data, γ_{2w} is the variance, and $\phi(w)$ is the phase of the transfer function estimated from the bispectrum. The ARMA(p,q) impulse response is generated by taking the inverse Fourier-transform of the transfer function.

3.3.6   ARMA system identification Method

Estimation of the MA(q) Parameters of the ARMA(p,q) System

The aim in using a residual time series approach to ARMA modelling was to exploit the method of moving average blind system identification, MN96, described earlier in this chapter. In order to exploit the ability of the algorithm

MN96 to identify MA processes with symmetrical input distributions, the method for estimation of the AR(p) parameters must also identify systems correctly for both symmetrical and asymmetrical input distributions.

Obviously, if the AR parameter estimates are incorrect then the estimation of the MA parameters is to some extent compromised. However, in producing an ARMA(p,q) method using a residual time series method the MA(q) must be resilient to such errors. During simulation using a variety of different filter configurations estimation of the MA parameters occasionally resulted in values of zero being returned. This failure occured for particularly poor AR(p) estimates and perturbation of the MA method did not result in a set of better MA(q) estimates. It must be stressed that this effect was very rare, but for cases where the MA(q) method MN96 resulted in a null set of MA(q) parameter estimates the results of the c(q,k) method used to initialise method MN96 were taken as the MA(q) parameter estimates. The method for AR parameter estimation is described below.

A Mixed-Order Method for Estimation of the AR(p) Parameters of an ARMA(p,q) System

GM89 is a blind system identification method which used a residual time series method. The estimation of the AR parameters $\{a(i)\}_{i=0}^{p}$ is performed outside the region of influence of the MA part of the process by using cumulants with lags greater than the MA model order, q. Equation (3.61) related the third-order cumulants to the AR parameter estimates. Equations relating second-order and fourth-order cumulants to the AR parameters are derived in an identical manner and the result in equations (3.73) and (3.75). The third-order relation is repeated here in equation (3.74).

$$\sum_{i=0}^{p} a(i)c_{2x}(m-i) = 0 \qquad m > q \tag{3.73}$$

$$\sum_{i=0}^{p} a(i)c_{3x}(m-i,n) = 0 \qquad m > q \tag{3.74}$$

$$\sum_{i=0}^{p} a(i)c_{4x}(m-i,n,s) = 0 \qquad m > q \tag{3.75}$$

These three separate equations each provide a method of AR parameter identification using the relevant order cumulant sequences. When additive

Gaussian noise is superimposed onto the noise-free output sequence the performance of the second-order method deteriorates whilst the higher-order methods achieve noise rejection such that performance is not compromised. However, there are disadvantages in using higher-order methods. In order to maintain results in terms of mean estimates and standarddeviation of estimates a longer data sequence is required for higher order methods than for second-order methods. By using a combination of these methods in a least squares approach the aim is to provide a generic system identification method that will perform well for the broadest range of systems irrespective of input distribution and additive noise.

Equations (3.73) - (3.75) are all equated to zero and could be summed to produce a mixed-order cumulant method. However, for any symmetrical input including a Gaussian input the use of the third-order cumulants would result in a poorer estimate of the system parameters. Alternatively, these equations could be summed using different scaling factors. This approach raises two issues: whether an optimal scaling vector can be found for generic rather than specific systems, and how the scaling vector could be formed to use only relevant cumulant orders for different input distributions or noise levels and types. In fact this approach would lead to system specific solutions rather than a generic algorithm. Instead, a method for deciding whether the r-th order cumulants of the output are defective due to a specific input distribution being used is proposed. The method eliminates all equations derived using those r-th order cumulant relations in an attempt to improve AR, and hence MA, parameter estimation.

Equations (3.73) - (3.75) are computed over a range of m, n, s and the singular value decomposition of each set of single order equations is taken. A decision is made whether or not to include the set of equations for that cumulant order upon the outcome of thresholding the singular values of each order cumulant equations. Thresholding is optimised to exclude third-order cumulants when the system is detected as having a symmetrical input distribution and third- and fourth-order cumulants if the input is detected as being Gaussian in distribution. From research on data sets generated for different ARMA(p,q) systems using symmetrical, asymmetrical and Gaussian input distributions a thresholding technique was chosen. The maximum singular value relating to the unmixed sets of equations in terms of second-, third- and fourth-order cumulants was extracted. If the r-th order maximum singular value was above 0.5 then the set of r-th order equations was included in the AR(p) calculation. Orders with a maximum singular value below 0.5 were rejected and the AR(p) parameters were estimated from the other cumulant-orders. This method of r-th order cumulant rejection is simple and effective. Table (3.9) gives results that are representative of all the filters simulated.

Table 3.9: Symmetrical and Gaussian Detection Rates

Input type	Laplacian		Gaussian		Exponential	
SNR dB	∞	10	∞	10	∞	10
detected symmetrical	89	85	4	3	0	0
detected Gaussian	5	8	94	92	0	0

The results relate to the NMP ARMA(3,2) filter whose coefficients are

$$a(0) = 1 \quad a(1) = -1.435 \quad a(2) = 0.886 \quad a(3) = -0.128$$

$$b(0) = 1 \quad b(1) = -1.3 \quad b(2) = -1.4$$

and the results are the number of times during 100 Monte Carlo runs that the simple thresholding technique detected the input distribution to be Gaussian or symmetrical at the signal-to-noise ratios specified using additive white Gaussian Noise (AWGN). The results of table (3.9) show that the simple thresholding technique satisfactorily detects Gaussianity and symmetry in the input distribution from the output cumulants only. That the input distribution type is decided from the statistics of the output data alone is essential and ensures that the ARMA(p,q) retains its blind identification status. Once the orders of cumulants to be used in the AR(p) parameter identification method have been established the overdetermined set of equations created from setting $m > Q$ (where Q is an over-estimate of the MA model order) in equations (3.73) - (3.75) is solved using least squares minimisation.

3.3.7 ARMA Model Order Selection

Model order selection in ARMA systems is a problem that has been addressed in a number of ways. The best known of these methods include the Final Prediction-Error (PDE) criterion and the Akaike Information Criterion (AIC) proposed by [1] which are asymptotically equivalent but do not yield consistent model order estimates. The Minimum Description Length (MDL) criterion proposed simultaneously by [40] and [38] has been shown to produce consistent model order estimates. [29] revised the method of [38] to produce a method for AR(p), as opposed to ARMA(p,q), model order selection. Methods based on the FPE, AIC and MDL criterion calculate the prediction-error variance over a range of (p, q) and select the model order to be the pairing of p and q that yields the smallest value of the selected criterion. Such approaches are computationally expensive because all the parameter estimates must be made prior to determination of the model order. [15] detailed a method of model order selection which is based on the MDL approach but which avoids the need for estimating all the model parameters by making use of the eigenvalues of the covariance matrix.

Alternatively model order selection in ARMA systems can be performed using an RTS type approach. Firstly, the AR(p) order is determined by any relevant means but most commonly from the rank of an extended cumulant matrix. The AR(p) parameters are then identified. Secondly, a residual time series is created by removing the influence of the AR(p) process from the original timeseries. Any MA order selection method can then be used to determine the MA model order, q, prior to the identification of the MA(q) parameters. This type of approach is suitable for use with all the ARMA(p,q) identification methods described here. Many methods of AR(p) model order selection have been proposed and some of these have been extended to deal with ARMA(p,q) model order selection. [7] compares the performance of various AR(p) model order selection methods for a variety of systems. It covers methods decribed by [1], [38], [14], [8] and [39].

3.3.8   Simulation Details

Filter Coefficients

M Monte Carlo runs of data length, K, were generated using function AR-MASYN.M of the HOSA toolbox of MATLAB with the AR(p) and MA(q) parameters of the specific filters defined in table (3.10) and with the input distribution specified for each example. Coloured or white Gaussian noise was then added to the noise-free filtered data to produce the signal-to-noise ratios detailed. All filters have $a(0) = b(0) = 1$. The phase of the filters is given in table (3.10) where NMP refers to Non-minimum phase and MP refers to minimum phase systems. Results used to compare parameter estimates using different ARMA(p,q) methods use identical data sets.

Cumulant Estimates

Cumulant estimates are performed using the method described in chapter 1 of this book. None of the more complicated methods which aim to improve robustness of estimates was implemented for any of the ARMA(p,q) identification methods.

Table 3.10: ARMA Filter Coefficients

filter	AR coefficients				MA coefficients				Phase
	a(0)	a(1)	a(2)	a(3)	b(0)	b(1)	b(2)	b(3)	
arma1	1	-0.3	-0.4	-	1	0.75	-2.5	-	NMP
arma2	1	-1.3	1.05	-0.325	1	-2.95	1.90	-	NMP
arma3	1	-1.5	1.21	-0.455	1	-1.5616	0.5358	-	NMP
arma4	1	-1.435	0.886	-0.128	1	-1.3	-1.4	-	NMP
arma5	1	0.4	-0.2	-0.6	1	-0.22	0.17	-0.1	MP
arma6	1	-1.4	0.95	-	1	-0.86	0.431	-	MP

Model Order

For all simulations the correct model orders of the filters were used. No estimation of the ease of model order selection for these filters has been made.

3.3.9 Method GM89

Using the residual time series method of [11] with a combination of different cumulant-orders gives single-order methods GM89-2, GM89-3, GM89-4, which use second-, third- and fourth-order cumulants respectively, and mixed-order methods GM89-23 and GM89-24 which use a combination of second- and third-order, and second- and fourth-order, cumulants respectively. Regardless of the orders of cumulants to be used the method first estimates the AR(p) parameters, then forms an AR-compensated residual time series from which it finally estimates the MA parameters. Function ARMARTS.M of MATLAB's HOSA toolbox is used to estimate the ARMA parameters of an ARMA(p,q) process where the orders p and q are explicitly given. Some adaptation of this function has been made to allow a second-order only method, GM89-2. All other methods employ an unadapted version of function ARMARTS.M. All simulations were made using unbiased sample estimates, no overlap between segments and 256 samples per segment. The maximum number of cumulant lags used was set to its default value of $(p+q)$.

Q-Slice Method

Function ARMAQS.M of the HOSA toolbox of MATLAB is a realisation of the q-slice method described by [12]. It allows either third- or fourth-order cumulants to be used in the parameter estimation. The use of third-order cumulants in simulations is denoted by method QS-3 whilst method QS-4 denotes the use of fourth-order cumulants in function ARMAQS.M. All simulations used unbiased sample estimates, no overlap between segments and 256 samples per segment. The maximum number of cumulant lags used was set to its default value of $(p + q)$.

Method DCQK

[12] proposed a method which has become known as the double c(q,k) method. When the AR model order is set to zero this method simplifies to an MA(q) method which was previously described by [10] and was termed the c(q,k) algorithm. This method is unusual in that it can identify noncausal systems in addition to causal systems. As such the results of [12] naturally reflect this feature of noncausal system identification. No results for causal systems are given. [25] does give results for simulations performed on a causal system but these are unsuitable for comparison of different methods because they employ a learning algorithm in addition to the double c(q,k) algorithm of [12]. In order to provide comparitive results an implementation of the method, labelled DCQK, was made. This implementation differs from the method described in [10] slightly. These differences are emphasised below. All results labelled DCQK relate to simulations performed using the author's implementation of the double c(q,k) algorithm.

Table (3.11) shows results for an ARMA(2,1) system simulated in [12]. The results labelled GS90 double c(q,k) are copied directly from the relevant table of [10], whilst the results labelled DCQK are from simulations undertaken using the author's implementation which meet the conditions of [10] as closely as possible. The results show that implementation DCQK performs to a similar level in terms of mean parameter estimates over 100 Monte Carlo runs using the conditions specified in [10], namely a data length of $K = 1024$, with additive white Gaussian noise at a signal-to-noise ratio $SNR = 20$dB. Differences in the mean estimates and standard deviations are attributed to incomplete knowledge of the full set of conditions under which the simulations whose results are given were performed. In particular, the impulse response used to create the output data on which simulations were performed was truncated to reflect figure 5 of [12] and the least squares approach was used. In addition the singular value decomposition, (SVD), method described in [12] was not implemented and results labeled DCQK use a least squares minimisation. This approach was chosen to allow comparison of different methods on the most similar terms possible. The SVD approach of [12] could be used for all the other methods and would be ex-

Table 3.11: Verification of Double c(q,k) Implementation DCQK

Method	a(1)	a(2)	b(1)
	-0.5	-0.9375	-0.5
GS90 double c(q,k)	-0.428 ± 0.164	-1.008 ± 0.215	-0.626 ± 0.727
DCQK	-0.382 ± 0.134	-0.956 ± 0.233	-0.598 ± 0.511

pected to improve the results. As such it was considered most appropriate to remove this advantage from the method and use a least squares minimisation of the equations. However the results for the row labeled GS90 double c(q,k) below are taken directly from table 2 of [12] and are the result of implementation of the SVD improvment. The results of method DCQK are therefore expected to be a little worse in terms of both mean estimate and standard deviation. The implementation was assumed verified and results for the double c(q,k) method presented in this chapter use this different implementation termed DCQK.

3.3.10   Discussion and Results

AR Parameter Estimation

Table (3.12) shows the mean estimated AR parameters of *arma4* using a noise-free data of length $K = 1024$ averaged over 30 Monte Carlo simulations and a zero-mean, negative exponential input distribution. Methods GM89-2, GM89-3, GM89-4, GM89-23 and GM89-24 use the HOSA toolbox function ARMARTS.M with the cumulant orders used set to second-order, third-order, fourth-order, second- and third-order, and second- and fourth-order respectively. Methods QS-3 and QS-4 use the Q-slice method implemented in the HOSA toolbox in ARMAQS.M and using third-order, and fourth-order cumulants respectively. Whist the performance is comparable for the GM89 methods, except for the method using fourth-order cumulants alone, the third-order QS method and method JKR96 in terms of mean estimates and standard deviations, method QS-4 fails to identify the system in any real sense.

Table (3.13) uses an identical data length averaged over the same number of Monte Carlo runs but with the input distribution set to a Laplacian distribution which is symmetrical. Notice that the performance of GM89-3, GM89-23 and QS-3 have been compromised by a symmetrical input distribution.

Table 3.12: Comparison of AR parameter estimates for *arma4*: Asymmetrical input distribution

Method	a(1)	a(2)	a(3)
	-1.435	0.886	-0.128
GM89-2	-1.488 ± 0.218	0.939 ± 0.238	-0.155 ± 0.132
GM89-3	-1.433 ± 0.203	0.870 ± 0.249	-0.124 ± 0.131
GM89-4	-1.169 ± 0.501	0.636 ± 0.577	-0.025 ± 0.301
GM89-23	-1.507 ± 0.138	0.963 ± 0.162	-0.171 ± 0.093
GM89-24	-1.415 ± 0.275	0.871 ± 0.296	-0.119 ± 0.161
QS-3	-1.598 ± 0.304	1.094 ± 0.393	-0.245 ± 0.208
QS-4	25.275 ± 150.983	-29.710 ± 174.890	15.958 ± 92.147
JKR96	-1.502 ± 0.159	0.957 ± 0.210	-0.169 ± 0.122

Table 3.13: Comparison of AR parameter estimates for *arma4*: Symmetrical input distribution

Method	a(1)	a(2)	a(3)
	-1.435	0.886	-0.128
GM89-2	-1.459 ± 0.231	0.980 ± 0.249	-0.138 ± 0.136
GM89-3	-1.573 ± 0.499	1.228 ± 0.833	-0.492 ± 0.741
GM89-4	-1.307 ± 0.374	0.795 ± 0.5323	-0.118 ± 0.326
GM89-23	-1.711 ± 0.268	1.216 ± 0.307	-0.307 ± 0.166
GM89-24	-1.475 ± 0.255	0.944 ± 0.293	-0.156 ± 0.165
QS-3	-2.911 ± 0.869	2.223 ± 0.984	-1.031 ± 0.543
QS-4	-3.244 ± 3.175	3.253 ± 4.645	-1.419 ± 2.718
JKR96	-1.448 ± 0.157	0.901 ± 0.199	-0.1343 ± 0.108

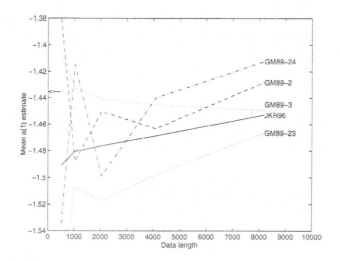

Figure 3.8: Estimates of a(1) for *arma4* using a Negative Exponential Input Distribution

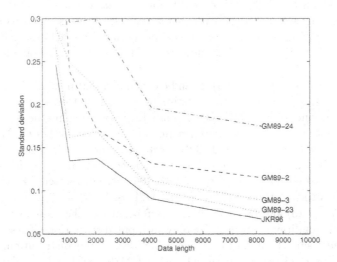

Figure 3.9: Standard Deviations of a(1) Estimates for *arma4* using a Negative Exponential Input Distribution

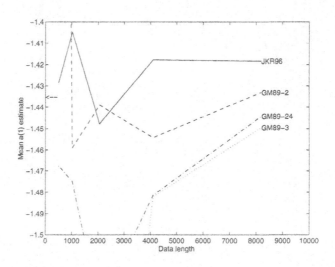

Figure 3.10: Estimates of a(1) for *arma4* using a Laplacian Input Distribution

Figures (3.8) and (3.9) show the performance of the different cumulant order methods of GM89 and of JKR96 for a negative exponential input distribtution, noise-free system with increasing data lengths in terms of estimated mean value of a(1) and standard deviation of the a(1) estimate. The true value of $a(1) = -1.435$ is indicated on figure (3.8) by an arrow at the y-axis.

From figures (3.8) and (3.9) it can be concluded that most of the methods perform relatively well. However, estimates for GM89-4 possessed a greater mean error and standard deviation than those of other methods. Due to requirements to resolve the performance of the other methods GM89-4 is omitted from figures (3.8) and (3.9). Note that the standard deviations of all methods decrease with increasing data length. In terms of both standard deviation and mean estimate method JKR96 performs better than either of the mixed-order methods, GM89-23 and GM89-24.

Figures (3.10) and (3.11) show the performance of the same methods for simulation conditions which differ only in input distribution. These figures reflect the result of simulations using a Laplacian input distribution. Note that the standard deviations of a(1) using GM89-3 with a Laplacian input distribution do not fall with increasing data length. This is indicative of the failure of the method to adequately determine the parameter estimate. Estimates GM89-23 and GM89-4 are omitted from figures (3.10) and (3.11).

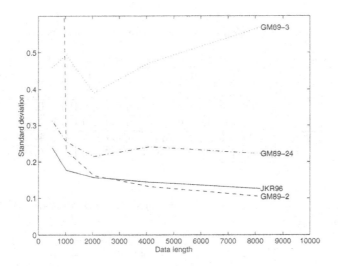

Figure 3.11: Standard Deviations of a(1) Estimates for *arma4* using a Laplacian Input Distribution

They produced mean estimates with a much greater error than the other methods and depiction would have resulted in the resolution of the other methods being lost.

As predicted by the theory, the third-order cumulant methods GM89-3 and QS-3 fail for symmetrical input distributions. Notice that JKR96 still outperforms mixed-order method GM89-24. Because the results are from simulations where the output is noise-free method GM89-2 would be expected to produce a mean estimate which is both less biased and exhibits less standard deviation than any higher-order method and this is the case. However, performance of method JKR96 is comparable even under conditions which are known to favour a second-order method.

Overall Performance

Table (3.14) shows results for the four methods using noise-free output data of length of $k = 2048$ averaged over 30 Monte Carlo runs with a negative exponential input distribution for filter arma6. Performance for this minimum phase filter is similar in terms of mean estimates and standard deviations of both AR and MA parameters for methods JKR96, GM89-23 and DCQK. Method QS-3 fails to determine either of the MA parameters satisfactorily due to the very large standard deviation over the 30 Monte Carlo simulations. Coloured Gaussian noise, $v[k]$, with

Table 3.14: Comparison of Performance for *arma6*

Method	a(1)	a(2)	b(1)	b(2)
	-1.4	0.95	-0.86	0.431
JKR96	-1.398 ± 0.021	0.951 ± 0.015	-0.881 ± 0.166	0.430 ± 0.049
GM89-23	-1.397 ± 0.020	0.948 ± 0.0173	-0.884 ± 0.170	0.429 ± 0.047
QS-3	-2.083 ± 0.798	0.877 ± 0.327	0.909 ± 7.795	-4.430 ± 13.665
DCQK	-1.333 ± 0.107	0.938 ± 0.099	-0.807 ± 0.127	0.431 ± 0.166

Table 3.15: Comparison of Performance for *arma6* with ACGN at SNR=10dB

Method	a(1)	a(2)	b(1)	b(2)
	-1.4	0.95	-0.86	0.431
JKR96	-1.395 ± 0.015	0.951 ± 0.018	-0.834 ± 0.117	0.428 ± 0.042
GM89-23	-1.392 ± 0.015	0.950 ± 0.012	-0.832 ± 0.112	0.428 ± 0.043
QS-3	-1.972 ± 0.449	0.836 ± 0.174	-0.233 ± 3.427	-1.525 ± 7.562
DCQK	-1.353 ± 0.075	0.930 ± 0.089	-0.806 ± 0.165	0.412 ± 0.178

$$v[k] = e[k] - 0.3e[k-1] + 0.2e[k-2] + 0.1v[k-1] - 0.15v[k-2]$$

and $e[k]$ being white Gaussian noise, was added to the noise-free signal to produce a signal-to-noise ratio SNR = 10dB. The performance of all four methods remained very similar to that shown for the noise-free case as shown in table (3.15). Filter *arma4* is a non-minimum phase system with poles located at $0.617 \pm j0.507$ and 0.200, and zeros located at 2 and -0.7. Table (3.16) shows results averaged over 30 Monte Carlo simulations, with a negative exponential input distribution being used, a data length of $k = 2048$ and white, Gaussian noise added to the noise-free output to give a noise-corrupted output signal with signal-to-noise ratio $SNR = 10$dB. Methods GM89-23, QS-3 and DCQK produce estimates for the MA(q) parameters which are unsatisfactory in terms of the standard deviation of the estimates over the 30 Monte Carlo runs. The mean parameter values shown in this table were used as the estimated filter coefficient values in figures (3.12) - (3.15). Table (3.17) shows the pole and zero locations of the estimated filters given in table (3.15) and in conjunction with figures (3.12) - (3.15) shows the effect of incorrect parameter estimates on pole and zero location and magnitude and phase characteristics. Figures (3.12) - (3.15) show how

Table 3.16: Estimated filter coefficients for *arma4*(3,2) using different identification techniques

Method	a(1)	a(2)	a(3)	b(1)	b(2)
	-1.435	0.886	-0.128	-1.3	-1.4
JKR96	-1.281 ± 0.206	0.717 ± 0.253	-0.041 ± 0.128	-0.650 ± 1.492	-0.788 ± 1.661
GM89-23	-1.369 ± 0.238	0.832 ± 0.298	-0.100 ± 0.154	11.906 ± 63.431	-2.508 ± 4.127
QS-3	-1.778 ± 0.596	1.349 ± 0.7656	-0.373 ± 0.388	-1.089 ± 3.177	1.842 ± 11.379
DCQK	-0.811 ± 0.141	0.104 ± 0.191	0.315 ± 0.149	1.822 ± 7.927	-1.513 ± 23.228

Table 3.17: Pole and Zero Locations for Estimated Filters

Method	Pole Locations		Zero locations	
	$0.617 \pm j0.507$	0.200	2	-0.7
JKR96	$0.609 \pm j0.520$	0.063	1.270	-0.620
GM89-23	$0.617 \pm j0.524$	0.156	-12.112	0.207
QS-3	$0.611 \pm j0.545$	0.556	$0.545 \pm j1.243$	
DCQK	$0.634 \pm j0.532$	-0.458	-2.442	0.620

effective the various methods have been in estimating the filter coefficients in terms of the resulting system magnitude and phase characteristics over the Nyquist frequency range. The true characteristics are shown by the solid line whilst the method indicated by the text below each part of the figure is shown by the dashed line. Table (3.18) gives results of simulations averaged over 30 Monte Carlo runs with a signal-to-noise ratio $SNR = 10$dB and with the output data length of $K = 2048$ and generated using a negative exponential input distribution filtered through a different NMP system, arma1. All the methods perform relatively well although the standard deviations for the DCQK method are noticeably higher. Results for table (3.19) use a Laplacian input distribution but with the remainder of the variables held constant for comparison.

Table (3.19) shows the effect of a symmetrical, Laplacian, input distribution on the performance of different methods. Methods GM89-23, QS-3 and DCQK all fail to identify the system correctly. Method JKR96 performs adequately, but with somewhat larger standard deviations. This is only to be expected since the method relies on the information carried by the second- and fourth-order cumulants in this case. Table (3.12) shows the relative performance of second-, third-, fourth-, second- and third-, and second- and fourth- order cumulants using the GM89 algorithm for a different system. Whilst this result is not for the same filter it is indicative of the general performance using the specified cumulant-orders. Notice that the standard deviation is always higher for the fourth- or mixed second- and fourth- order methods than for other cumulant order combinations. This result is reflected in the estimates using JKR96 for the symmetrical input distribution case compared to the use of a negative exponential input distribution.

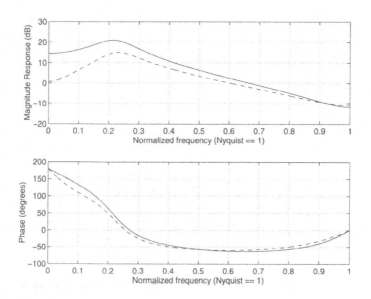

Figure 3.12: Comparison of Real (solid line) and Estimated (dashed line) Frequency Responses for JKR96

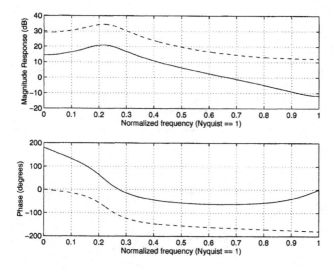

Figure 3.13: Comparison of Real (solid line) and Estimated (dashed line) Frequency Responses for GM89-23

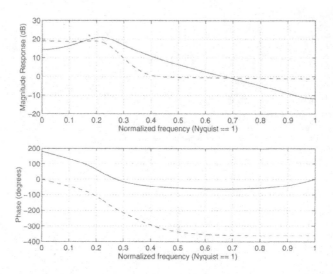

Figure 3.14: Comparison of Real (solid line) and Estimated (dashed line) Frequency Responses for QS-3

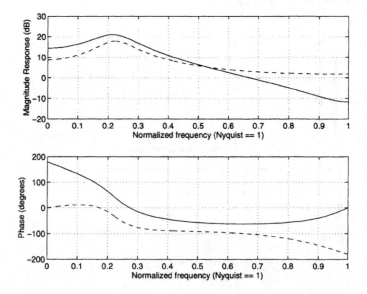

Figure 3.15: Comparison of Real (solid line) and Estimated (dashed line) Frequency Responses for DCQK

Table 3.18: Relative performance of JKR96, GM89-23, QS-3 and DCQK for negative exponential input distribution

Method	a(1)	a(2)	b(1)	b(2)
	-0.3	-0.4	0.75	-2.5
JKR96	-0.276 ± 0.048	-0.394 ± 0.061	0.875 ± 0.618	-2.563 ± 1.573
GM89-23	-0.293 ± 0.050	-0.400 ± 0.050	0.608 ± 0.493	-2.305 ± 1.209
QS-3	-0.058 ± 0.084	-0.486 ± 0.047	0.600 ± 0.145	-2.331 ± 0.362
DCQK	-0.186 ± 0.135	-0.337 ± 0.187	0.375 ± 1.644	-2.899 ± 9.417

Table 3.19: Failure of GM89-23, QS-3 and DCQK for Laplacian Input Distribution

Method	a(1)	a(2)	b(1)	b(2)
	-0.3	-0.4	0.75	-2.5
JKR96	-0.511 ± 0.396	-0.380 ± 0.827	0.348 ± 1.696	-1.012 ± 2.080
GM89-23	-0.460 ± 0.402	0.015 ± 0.375	6.064 ± 30.125	0.013 ± 0.872
QS-3	0.116 ± 1.513	0.145 ± 1.266	0.032 ± 3.853	-2.403 ± 12.435
DCQK	-0.035 ± 0.224	0.068 ± 0.275	1.485 ± 6.812	0.927 ± 4.076

3.3.11   Conclusions

Results for methods GM89-2, GM89-3, GM89-4, GM89-23, GM89-24, QS-3, QS-4 and JKR96 in terms of AR(p) parameter estimation for a range of filters under differing simulation conditions were presented. For the noise-free case with a negative exponential input distribution, methods GM89-2, GM89-3, GM89-23, GM89-24, QS-3 and JKR96 were shown to have similar levels of performance. However, for a Laplacian (symmetric) input distribution only methods GM89-2, GM89-24 and JKR96 performed adequately. The ability of method MN96, the MA(q) part of the ARMA(p,q) method JKR96, to function for any input distribution has already been proven in an earlier part of this chapter. Hence method JKR96 is shown to be the only higher order method suitable for blind system identification when no assumptions about the input distribution of the system can be made.

Methods GM89-23 and GM89-3 were shown to outperform the other GM89 methods in the case of an asymmetrical input distribution. Neither of these methods is suitable for identification of a system with a symmetrical input distribution. In order to show why a method such as JKR96, which will perform adequate blind system identification for any non-Gaussian input distribution, is needed method GM89-23 was chosen as the RTS method of [11] that would be simulated. It is understood that better performance for a specific case of input distribution may be possible using a different mixed-order cumulant method but a generic method of blind system identification was the aim of this research and the results are therefore presented to show how method JKR96 meets this aim whilst other methods fail. Also this approach will show method JKR96 against the best of the GM89 methods for all asymmetrical input distribution simulations. Method QS-4 performed poorly for both asymmetrical and symmetrical input distributions. The third-order cumulant implementation of the q-slice method, QS-3, was therefore chosen for comparison under full simulation.

Methods employing fourth-order cumulants were shown to exhibit greater bias and variance than methods using second- and third-order cumulants. However, these fourth-order methods were important in the identification of systems when the input distribution is symmetrical and the third-order cumulants are theoretically identically zero. Higher-order techniques performed to a similar standard as second order techniques for noise-free conditions and rapidly outperformed second-order methods when the observed output was corrupted by either coloured or white additive Gaussian noise. Higher-order methods can detect both phase and magnitude responses correctly and therefore nonminimum phase systems can be correctly identified. The best a second-order method of system identification can achieve for nonminimum phase systems is to identify a spectrally equivalent minimum phase model (SEMP) which has an identical magnitude response to the NMP system but exhibits different phase characteristics.

In terms of overall results for AR(p) and MA(q) parameter estimation method JKR96 significantly outperformed methods GM89-23, QS-3 and DCQK (which was implemented using third-order cumulants) or performed to a similar level (for example with minimum phase filter *arma6* where methods GM89-23, DCQK and JKR96 could not be separated in terms of overall performance for a negative exponential input distribution). Results for a symmetrical input distribution show that method JKR96 does indeed perform blind system identification for any non-Gaussian input distribution whilst methods GM89-23, QS-3 and DCQK do not work well enouh.

References

[1] H. Akaike. A new look at the statistical model identification. *IEEE Transactions Automatic Control*, AC-19:716 – 723, 1974.

[2] S. A. Alshebeili, A. N. Venetsanopoulos, and A. E. Cetin. Cumulant based identification approaches for nonminimum phase FIR systems. *Proceedings of IEEE*, 41:1576 – 1588, 1993.

[3] H. Bartelt, A. W. Lohman, and B. Wirnitzer. Phase and amplitude recovery from bispectra. *Applied Optics*, 23:3121–3129, 1984.

[4] M. Boumahdi. Blind identification using the kurtosis with applications to field data. *Signal Processing*, 48:205 – 216, 1996.

[5] D. R. Brillinger. The identification of a particular nonlinear time series system. *Biometrika*, 64:509–515, 1977.

[6] T. W. S. Chow and G. Fei. On the identification of non-minimum phase non-gaussian MA and ARMA models using a third-order cumulant. *International Journal of Electronics*, 79:839–852, 1995.

[7] J. R. Dickie and A. K. Nandi. A comparative study of AR order selection methods. *Signal Processing*, 40:239 – 255, 1994.

[8] P. M. Djuric and S. M. Kay. Order selection of autoregressive models. *IEEE Transactions on Signal Processing*, 40:2829 – 2833, 1992.

[9] J. A. Fonollosa and J. Vidal. System identification using a linear combination of cumulant slices. *Proceedings of IEEE*, 41:2405 – 2411, 1993.

[10] G. B. Giannakis. Cumulants: A powerful tool in signal processing. *Proceedings of IEEE*, 75:1333 – 1334, 1987.

[11] G. B. Giannakis and J. M. Mendel. Identification of nonmimimum phase systems using higher order statistics. *IEEE Transactions on Accoustics, Speech and Signal Processing*, 37:360 – 377, 1989.

[12] G. B. Giannakis and A. Swami. On estimating noncausal nonminimum phase ARMA models of non-gaussian processes. *IEEE Transactions on Accoustics, Speech and Signal Processing*, 38(3):478–495, March 1990.

[13] S. M. Kay. *Modern Spectral Estimation*. Prentice Hall, Englewood Cliffs, New Jersey, 1988.

[14] K. Konstantinides. Threshold bounds in svd and a new iterative algorithm for order selection in ar models. *IEEE Transactions on Signal Processing*, 39:757 – 763, 1991.

[15] G. Liang, D. M. Wilkes, and J. A. Cadzow. ARMA model order estimation based on the eignevalues of the covariance matrix. *IEEE Transactions on Signal Processing*, 41:3003 – 3009, 1993.

[16] K. S. Lii. Non-gaussian ARMA model identification and estimation. *Proc Bus and Econ Statistics (ASA)*, pages 135–141, 1982.

[17] K. S. Lii and M. Rosenblatt. Deconvolution and estimation of transfer function phase and coefficients for non-gaussian linear processes. *Annals of Statistics*, 10:1195–1208, 1982.

[18] K. S. Lii and M. Rosenblatt. Non-gaussian linear processes, phase and deconvolution. *Statistical Signal Processing*, pages 51–58, 1984.

[19] K. S. Lii and M. Rosenblatt. A fourth-order deconvolution technique for non-gaussian linear processes. In P. R. Krishnaiah, editor, *Multivariate Analysis VI*. Elsevier, Amsterdam, The Netherlands, 1985.

[20] P. Lin and S. Mao. Non-gaussian ARMA identification via higher order cumulant. *Signal Processing*, 33:357 –362, 1993.

[21] D. Mampel, A. K. Nandi, and K. Schelhorn. Unified approach to trimmed mean estimation and its application to bispectrum of eeg signals. *Journal of the Franklin Institute*, 333B:369 – 383, 1996.

[22] S. L. J. Marple. *Digital Spectral Analysis with Applications*. Prentice Hall, Englewood Cliffs, New Jersey, 1987.

[23] J. K. Martin and A. K. Nandi. Blind system identification using second, third and fourth order cumulants. *Journal of the Franklin Institute of Science*, 333B:1 – 13, 1996.

[24] T. Matsuoka and T. J. Ulrych. Phase estimation using the bispectrum. *Proceedings of IEEE*, 72:1403–1411, 1984.

[25] J. M. Mendel. Tutorial on higher-order statistics (spectra) in signal processing and system theory: Theoretical results and some applications. *Proceedings of IEEE*, 79:277 – 305, 1991.

[26] Y. J. Na, K. S. Kim, I. Song, and T. Kim. Identification of nonminimum phase FIR systems using third- and fourth-order cumulants. *IEEE
Transactions on Signal Processing*, 43:2018 – 2022, 1995.

[27] A. K. Nandi. On the robust estimation of third-order cumulants in applications of higher-order statistics. *Proceedings of IEE, Part F*, 140:380
– 389, 1993.

[28] A. K. Nandi. Blind identification of FIR systems using third order
cumulants. *Signal Processing*, 39:131 – 147, 1994.

[29] A. K. Nandi and J. A. Chambers. New lattice realisation of the predictive least-squares order selection criterion. *IEE Proceedings F*, 138:545
– 550, 1991.

[30] A. K. Nandi and D. Mampel. Improved estimation of third order cumulants. *FREQUENZ*, 49:156 – 160, 1995.

[31] A. K. Nandi and R. Mehlan. Parameter estimation and phase reconstruction of moving average processes using third order cumulants. *Mechanical Systems and Signal Processing*, 8:421 – 436, 1994.

[32] C. L. Nikias. ARMA bispectrum approach to nonminimum phase system identification. *IEEE Transactions on Accoustics, Speech and Signal
Processing*, 36:513 – 524, 1988.

[33] C. L. Nikias. Higher-order spectral analysis. In S. S. Haykin, editor, *Advances in Spectrum Analysis and Array Processing*, volume I, chapter 7.
Prentice Hall, Englewood Cliffs, New Jersey, 1991.

[34] C. L. Nikias and H. H. Chiang. Higher-order spectrum estimation via
noncausal autoregressive modeling and deconvolution. *IEEE Transactions on Accoustics, Speech and Signal Processing*, 36:1911 – 1913, 1988.

[35] C. L. Nikias and J. M. Mendel. Signal procesing with higher-order
spectra. *IEEE Signal Processing Magazine*, pages 10 – 37, 1993.

[36] C. L. Nikias and R. Pan. ARMA modelling of fourth-order cumulants and phase estimation. *Circuits, Systems and Signal Processing*,
7(13):291–325, 1988.

[37] J. K. Richardson. *Parametric modelling for linear system identification
and chaotic system noise reduction*. PhD thesis, University of Strathclyde, Glasgow, UK, 1996.

[38] J. Rissanen. Modeling by shortest data description. *Automatica*, 14:465 – 471, 1978.

[39] J. Rissanen. A predictive least-squares principle. *I M A J. Math. Control Inform.*, 3:211 – 222, 1986.

[40] G. Schwarz. Estimation of the dimension of a model. *Annals of Statistics*, 6:461 – 464, 1978.

[41] L. Srinivas and K. V. S. Hari. FIR system identification using higher order cumulants - a generalised approach. *IEEE Transaction on Signal Processing*, 43:3061 – 3065, 1995.

[42] A. G. Stogioglou and S. McLaughlin. MA parameter estimation and cumulant enhancement. *IEEE Transactions on Signal Processing*, 44:1704 –1718, 1996.

[43] A. Swami and J. M. Mendel. ARMA parmaeter estimation using only output cumulants. *IEEE Transactions on Accoustics, Speech and Signal Processing*, 38:1257 – 1265, 1990.

[44] I. The MathWorks. *HOSA toolbox for use with MATLAB*.

[45] J. K. Tugnait. Identification of non-minimum phase linear stochastic systems. *Automatica*, 22:457–464, 1986.

[46] J. K. Tugnait. Identification of linear stochastic systems via second- and fourth-order cumulant matching. *IEEE Transactions on Information Theory*, 33:393–407, 1987.

[47] J. K. Tugnait. Approaches to FIR system identification with noisy data using higher order statistics. *IEEE Transactions on Accoustics, Speech and Signal Processing*, 38:1307 – 1317, 1990.

[48] J. K. Tugnait. New results on FIR system identification using higher order statistics. *IEEE Transactions on Signal Processing*, 39:2216 – 2221, 1991.

[49] X. Zhang and Y. Zhang. FIR system identification using higher order statistics alone. *IEEE Transactions on Signal Processing*, 42:2854 – 2858, 1994.

[50] Y. Zhang, D. Hatzinakos, and A. N. Venetsanopoulos. Bootstrapping techniques in the estimation of higher-order cumulants from short data records. *Proceedings of the International Conference of Accoustics, Speech and Signal Processing*, IV:200 – 203, 1993.

4 Blind Source Separation

V Zarzoso and A K Nandi

Contents

4.1 Introduction 169

4.2 Problem statement 171

    4.2.1 General BSS problem. Convolutive and instantaneous linear mixtures 171

    4.2.2 Assumptions 174

    4.2.3 Waveform-preserving solutions 177

    4.2.4 General approach. Identifiability 178

    4.2.5 Geometrical insights 179

    4.2.6 An algebraic standpoint: cumulant tensor diagonalization 180

4.3 Separation quality: performance indices 185

    4.3.1 Measure of the source-waveform preservation 185

    4.3.2 Quality of the estimated mixing matrix 186

    4.3.3 Interference-to-signal ratio 186

    4.3.4 Degree of statistical independence 187

    4.3.5 Computational complexity 187

4.4 A real-life problem: the fetal ECG extraction 188

    4.4.1 BSS model of the FECG extraction problem 188

    4.4.2 A real 8-channel ECG recording 189

4.5 Methods based on second-order statistics 189

    4.5.1 Principal component analysis (PCA) 191

        PCA via EVD 192

        PCA via SVD 192

    4.5.2 Justifying HOS: indeterminacy of second-order analysis 195

      4.5.3   Examples 199

4.6   Methods based on higher-order statistics 203

      4.6.1   Higher-order eigenvalue decomposition (HOEVD) 203

             Contrast functions 203

             Pairwise versus mutual independence 206

             Extension to general BSS set-up 207

             Links and remarks 208

             Examples 209

      4.6.2   Higher-order singular value decomposition (HOSVD) . . 211

             Higher-order arrays and the HOSVD 211

             The HOSVD method 214

             Examples 216

      4.6.3   Extended Maximum-Likelihood (EML) 217

             Preliminaries 217

             A rotation angle estimator 220

             Connection with other methods 222

             Geometrical interpretation 223

             Examples 224

      4.6.4   Equivariant source separation 226

             Equivariant adaptive algorithms 228

             Examples 230

4.7   Comparison 231

      4.7.1   Experimental methodology 233

      4.7.2   Results and discussion 234

4.8   Comments on the literature 236

References 247

4.1 Introduction

A myriad of applications require the extraction of a set of signals which are not directly accessible. Instead, this extraction must be carried out from another set of measurements which were generated as mixtures of the initial set. Since usually neither the original signals — called *sources* — nor the mixing transformation are known, this is certainly a challenging problem of multichannel blind estimation. One of the most typical examples is the so-called "cocktail party" problem. In this situation, any person attending the party can hear the speech of the speaker they want to listen to, together with surrounding sounds coming from other 'competing' speakers, music, background noises, etc. Everybody has experienced how the human brain is able to separate all these incoming sound signals and to 'switch' to the desired one. Similar results can be achieved by adequately processing the output signals of an array of microphones, as long as the signals to be extracted fulfil certain conditions [62, 63]. Wireless communications is another usual application field of signal separation techniques. In a CDMA (Code Division Multiple Access) environment several users share the same radio channel by transmitting their signal after modifying it according to an appropriate code. Traditionally, the extraction of the desired signal at the receiving end requires the knowledge of the corresponding code. However, if an antenna array is available, each of its elements would pick up a mixture of the transmitted coded signals, and then they could be recovered by applying blind separation procedures. The interesting fact is that no prior decoding process is necessary in their recovery if BSS techniques are employed. Again, for this to hold true some assumptions must be met by the system. Blind separation of sources proves also useful in biomedical applications. The separation of the maternal and the fetal electrocardiograms is one of them. This problem arises when the fetus' heart condition is to be monitored during pregnancy by means of non-invasive techniques, that is, by attaching electrodes to several points on the mother's skin. The electrodes output a mixture of mother's and fetus' heartbeat signals, apart from some other disturbances such as thermal noise, mains interference, maternal electromyogram interference, etc. For diagnostic purposes, it is desired to extract the fetal heart contributions from the corrupted cutaneous recordings. This task can be accomplished by blind separation methods [2, 8, 38, 46, 61, 68]. Other applications comprise many diverse areas such as [18, 19, 21, 24, 31, 59, 62–64]:

- *Array signal processing*: separation and recognition of sources from unknown arrays, direction of arrival (DOA) estimation, source localization from perturbed arrays.

- *Radar and sonar*: jammer rejection, radar and sonar detection, airport surveillance, beamforming.

- *Communications*: mobile radio and regenerative satellite communications, data communications in the presence of cross-coupling effects.

- *Speech processing*: speech recorded in the presence of background noise and/or competing speakers, automatic voice recognition in noisy acoustic environments.

- *Medicine*: determination of firing patterns of neural signals from electromyograms (EMG) measured from skin electrodes, elimination of electrocardiogram (ECG) artifacts, separation and determination of brain activity sources.

- *Semiconductor manufacturing*: monitoring the status of the key process parameters (e.g., diffusion times, times and temperatures, gas fluxes, etc.) from the process testing data (e.g., threshold voltage, drive current, sheet resistance, parasitic capacitance, etc.)

- *Circuit testing and diagnosis*: circuit temperature monitoring in high power-density electronic circuits, determination of randomly located input signals from output signals.

- *Image processing*: reconstruction and restoration of distorted images.

- *Statistics*: factor analysis, data pre-processing before Bayesian detection and classification, projection pursuit.

With so many practical applications, it is no wonder that the problem of the blind separation of sources has aroused such enormous research interest among the signal processing community since the mid-eighties.

The main goal of the present chapter is to supply the reader with an insight into the blind source separation (BSS) problem and its basic foundations, as well as to summarize the rationale behind a number of methods for BSS. In the first place, the BSS problem is endowed with a general mathematical formulation. We will focus on real instantaneous linear mixtures of real signals, since they condense the spirit of the BSS. Nevertheless, the extension to complex-valued mixtures is often straightforward. Then the mathematical and statistical grounds to cope with the BSS problem are introduced. Although it seems ill-posed, the extraction of the sources is possible under certain assumptions and identifiability conditions, which are also highlighted. After presenting some parameters helpful in measuring the separation performance, several methods for BSS are studied. Basically, they can be split into two main groups. The first group consists of procedures based on second-order statistics (SOS) alone and are known as *principal component analysis*. Their major advantage is their simplicity, attained at the expense of an indeterminacy in the results for linear mixtures. The other group of BSS methods are generally referred to as *independent component analysis*. Apart from second-order analysis, they further resort to the higher-order statistics (HOS) of the available data. This results in more elaborate algorithms, but the indeterminacy vanishes and the separation performance is remarkably enhanced relative to the second-order procedures. All the methods presented herein are illustrated throughout the chapter with the aid of some experimental examples and an application involving real signals: the extraction of the fetal ECG from electrode recordings taken from the mother's skin. A commentary on some of the existing literature on the subject brings the chapter to an end.

4.2 Problem statement

4.2.1 General BSS problem. Convolutive and instantaneous linear mixtures

Consider the set of q **source signals** $\{x_i(t),\ i = 1, 2, \ldots, q\}$ which generate the set of p **observations** or **measurements** $\{y_i(t),\ i = 1, 2, \ldots, p\}$ at the sensor output, where t represents a time index. In the general case, the measurements can be regarded as mixtures of transformed versions of the sources, contaminated by some additive **noise** $\{n_i(t),\ i = 1, 2, \ldots, p\}$:

$$y_i(t) = \sum_{j=1}^{q} H_{ij}\{x_j(t)\} + n_i(t), \qquad i = 1, 2, \ldots, p, \qquad (4.1)$$

where $H_{ij}\{\cdot\}$ denotes the transformation carried out on the jth source contributing to the ith sensor signal. In the linear case, $H_{ij}\{\cdot\}$ is a linear time-invariant system operating over $x_j(t)$, $H_{ij}\{x_j(t)\} = h_{ij}(t) * x_j(t)$, symbol '$*$' denoting the convolution operator, and then the ith measurement becomes:

$$y_i(t) = \sum_{j=1}^{q} h_{ij}(t) * x_j(t) + n_i(t), \qquad i = 1, 2, \ldots, p. \qquad (4.2)$$

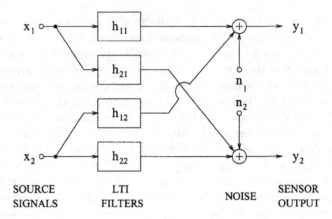

Figure 4.1: Convolutive signal separation model for the two-source two-sensor case.

This is called the **convolutive mixture BSS model** [62–64]. Abusing notation, this model can also be written in matrix form as:

$$
\begin{bmatrix} y_1(t) \\ \vdots \\ y_p(t) \end{bmatrix} = \begin{bmatrix} h_{11}(t) & \cdots & h_{1q}(t) \\ \vdots & \ddots & \vdots \\ h_{p1}(t) & \cdots & h_{pq}(t) \end{bmatrix} * \begin{bmatrix} x_1(t) \\ \vdots \\ x_q(t) \end{bmatrix} + \begin{bmatrix} n_1(t) \\ \vdots \\ n_p(t) \end{bmatrix} \Rightarrow
$$

$$
\Rightarrow \quad \mathbf{y}(t) = H(t) * \mathbf{x}(t) + \mathbf{n}(t), \quad (4.3)
$$

where the column vectors $\mathbf{y}(t)$, $\mathbf{x}(t)$ and $\mathbf{n}(t)$ contain the observations, the sources and the noise signals, respectively. Matrix $H(t)$ represents the multi-input multi-output linear system (channel) which links sources with measurements. The matrix notation above stresses the multi-channel nature of the problem. Figure 4.1 graphically depicts the BSS model for convolutive mixtures, in the simplified scenario of two sources and two measurements.

In many applications this model can be further simplified by assuming that the linear filters $h_{ij}(t)$ are composed of a single coefficient: $h_{ij}(t) = m_{ij}\delta(t)$. This occurs in narrowband propagation conditions, for instance, where the propagation delays from sources to sensors can be reduced to phase delays. As a result, the propagation effects are reduced to a change in amplitude, $|m_{ij}|$, and a phase term, $e^{j\angle m_{ij}}$. Under these circumstances,

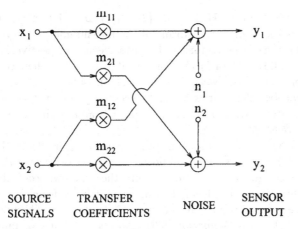

SOURCE   TRANSFER   SENSOR
SIGNALS   COEFFICIENTS   NOISE   OUTPUT

Figure 4.2: Instantaneous linear mixture model for two sources and two measurements.

equation (4.2) becomes:

$$y_i(t) = \sum_{j=1}^{q} m_{ij} x_j(t) + n_i(t), \qquad i = 1, 2, \ldots, p, \qquad (4.4)$$

or, in matrix form:

$$\begin{bmatrix} y_1(t) \\ \vdots \\ y_p(t) \end{bmatrix} = \begin{bmatrix} m_{11} & \cdots & m_{1q} \\ \vdots & \ddots & \vdots \\ m_{p1} & \cdots & m_{pq} \end{bmatrix} \begin{bmatrix} x_1(t) \\ \vdots \\ x_q(t) \end{bmatrix} + \begin{bmatrix} n_1(t) \\ \vdots \\ n_p(t) \end{bmatrix} \Rightarrow$$

$$\Rightarrow \quad \mathbf{y}(t) = M\mathbf{x}(t) + \mathbf{n}(t). \quad (4.5)$$

This is the **instantaneous linear mixture BSS model**. Now the matrix M contains the coefficients of the linear transformation which represents the transfer from sources to observations. This matrix is denoted as **mixing**, **transfer** or **channel-parameter matrix**. Figure 4.2 shows the instantaneous linear mixture model for the blind separation of two sources from two sensor outputs.

It is interesting to note that Fourier-transforming both sides of (4.3) yields in the frequency domain:

$$\mathbf{Y}(\omega) = \mathbf{H}(\omega)\mathbf{X}(\omega) + \mathbf{N}(\omega), \qquad (4.6)$$

where $\mathbf{Y}(\omega) = \mathrm{FT}\{\mathbf{y}(t)\}$, $\mathbf{H}(\omega) = \mathrm{FT}\{H(t)\}$, $\mathbf{X}(\omega) = \mathrm{FT}\{\mathbf{x}(t)\}$ and $\mathbf{N}(\omega) = \mathrm{FT}\{\mathbf{n}(t)\}$ are the Fourier transforms of the measurement vector, the multi-channel filter, the source vector and the noise vector, respectively. Therefore, both models (4.3) and (4.5) are somewhat equivalent when contemplated from different domains. Note, however, that in (4.6) the coefficients of the corresponding transfer matrix are not necessarily constant as in (4.5), but they may well vary with ω. References [62–64] address in detail the convolutive mixture case.

We will focus our attention on instantaneous linear mixtures. In this case, the goal of BSS can be stated as follows: to extract or recover the unknown original source signals, $\mathbf{x}(t)$, and possibly the unknown coefficients of the linear transformation that form the mixing matrix M from the only knowledge of the measured signals at the sensor output, $\mathbf{y}(t)$. The first objective (so-called *blind signal estimation* [58]) involves, explicitly or implicitly, the estimation of the transfer matrix (so-called *blind channel identification* [58]), so both goals are inextricably intertwinned. Figure 4.3 graphically illustrates the basic BSS set-up. As commented in the introduction, hereafter only real-valued mixtures will be considered, although often the extension to the complex case is relatively simple.

The term 'blind' highlights the fact that no a priori assumptions on the structure of the mixing matrix and the source signals are made, in contrast to other array processing techniques. For instance, in direction of arrival estimation, the transfer matrix is parameterized as a function of the arrival angles and the array layout [53, 55]. No such parameterization is done in BSS problems. And precisely there lies the versatility of the BSS model, since in many cases it is extremely difficult to model the transfer between sources and sensors or, simply, no *a priori* information is available about the mixture. In 'informed' array processing, deviations of the parameterized model from the actual array layout (so-called calibration errors) lead to errors in the estimated sources, which sometimes are difficult to quantify. No such deviations exist in the BSS model because nothing is assumed about the mixing structure. In the context of the beamforming problem, for instance, more advantages of blind techniques over 'informed' techniques are pointed out in [18].

4.2.2 Assumptions

A problem characterized by a model like (4.5) would be certainly ill-posed if no further assumptions were made about the characteristics of the system. These hypotheses can be divided into three groups, depending on whether they are related to the mixing matrix, the sources or the noise signals. Most

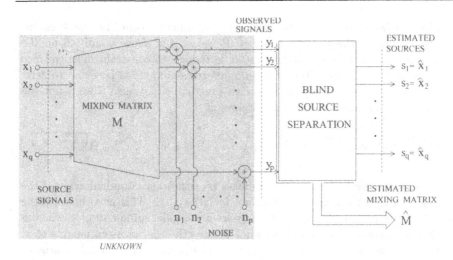

Figure 4.3: General blind source separation set-up.

methods for BSS rely on the following ones:

A1. The mixing matrix M is full column rank, with $p \geqslant q$ (i.e., number of sensors $\geqslant$ number of sources).

A2. Each component of $\mathbf{x}(t)$, $x_i(t)$, is a stationary zero-mean stochastic process, $i = 1, 2, \ldots, q$.

A3. At each time instant t, the components of $\mathbf{x}(t)$ are mutually statistically independent.

A4. The components of $\mathbf{x}(t)$ are unit-power.

A5. Each component of $\mathbf{n}(t)$, $n_i(t)$, is a stationary zero-mean process, $i = 1, 2, \ldots, p$.

A6. At each time instant t, the components of $\mathbf{n}(t)$ are mutually independent, as well as independent of the sources.

Assumption A3, also known as source *spatial whiteness* or *spatial independence*, is the crux of the blind separation. Statistically, it implies that the source cross-cumulants are null [7, 44, 47]:

$$\text{Cum}[x_{i_1}, \ldots, x_{i_r}] = 0, \qquad \forall (i_1, \ldots, i_r) \text{ not all equal}, \forall r \in \mathbb{N}. \qquad (4.7)$$

Physical independence of the sources is a sufficient condition for their independence in the statistical sense. However, the above condition on the source cumulants is too strong and it is enough to constrain independence up to order r, that is, that all sth-order cross-cumulants be equal to zero, for $s \leqslant r$, which is acceptable in most practical situations. If we define the rth-order cumulant array of the sources as

$$C^x_{i_1\ldots i_r} \triangleq \mathrm{Cum}[x_{i_1}, \ldots, x_{i_r}], \qquad 1 \leqslant i_1, \ldots, i_r \leqslant q, \qquad C^x_r \in \mathbb{R}^{\overbrace{q \times \cdots \times q}^{r}}, \tag{4.8}$$

then independence up to order r makes the sth-order cumulant array of the sources exhibit a diagonal structure, for all $s \leqslant r$. This is so because the diagonal entries of C^x_r contain the rth-order marginal cumulants of $\mathbf{x}$, whereas the off-diagonal elements correspond to the rth-order cross-cumulants of $\mathbf{x}$. In particular, at second order C^x_2 is equal to R_x, the source spatial covariance matrix:

$$R_x \triangleq \mathrm{E}[\mathbf{x}\mathbf{x}^T], \tag{4.9}$$

and thus R_x is also diagonal since the source cross-correlations (2nd-order cross-cumulants) are all null. This is just *second-order* (as opposed to order greater than two, or higher-order) *spatial whiteness* or *spatial uncorrelation*.

Assumption A4 is only a normalization convention. Actually the power of the sources can be included in the columns of M, leaving the observations unaltered. This property lies in the multiplicative-group structure of the model, which makes left-multiplication of the sources by a matrix equivalent to right-multiplication of the mixing matrix by the same linear transformation (associativity). Let P_i be the power of the ith source, $1 \leqslant i \leqslant q$. If $\mathbf{x}(t)$ has not unit-power signals, the equivalent set of sources

$$\tilde{\mathbf{x}}(t) = R_x^{-\frac{1}{2}}\mathbf{x}(t), \qquad R_x^{-\frac{1}{2}} = \mathrm{diag}\left(\frac{1}{\sqrt{P_1}}, \ldots, \frac{1}{\sqrt{P_q}}\right), \tag{4.10}$$

is such that $R_{\tilde{x}} \triangleq \mathrm{E}[\tilde{\mathbf{x}}\tilde{\mathbf{x}}^T] = I_q$, $q \times q$ identity matrix. Then the new equivalent mixing matrix is transformed according to:

$$M\mathbf{x}(t) = M(R_x^{-\frac{1}{2}}\tilde{\mathbf{x}}(t)) = (MR_x^{-\frac{1}{2}})\tilde{\mathbf{x}}(t), \tag{4.11}$$

so one can always think of the sources being unit-power by properly scaling the columns of the mixing matrix. Hence one can assume that

$$R_x = I_q \tag{4.12}$$

with no loss of generality whatsoever.

4.2.3 Waveform-preserving solutions

At this point, it is worth noticing the following technical difficulties concerning the determination of the sources [18,60]:

1. As commented previously, scalar factors can be exchanged between each source and the corresponding column of M without modifying the observations at all. That is, even bearing in mind the restriction imposed by assumption A4, the sign (in the real case) or phase (in the complex case) of each source signal remains unobservable.

2. The outputs of the separator cannot be ordered. As a matter of fact, the order of the sources is itself arbitrary, since arranging the sources and the columns of the mixing matrix accordingly leave the observations unchanged. As a conclusion, the source signals can at best be recovered up to a permutation.

However, these degrees of freedom in the sources that can be estimated at the output of any unmixing system do not actually represent a limitation in the separation process, because in most practical applications:

1. The information conveyed by the sources is contained in their waveforms, rather than in their magnitudes, and hence a finite scalar factor is acceptable.

2. The order in which the signals are arranged does not matter. An order change is thus acceptable.

Therefore, it is acceptable to obtain at the output of the separator a solution close to:

$$\hat{\mathbf{x}}(t) = \mathbf{s}(t) = PD\,\mathbf{x}(t), \tag{4.13}$$

where P is a ($q \times q$) permutation matrix and D a non-singular diagonal matrix of the same size. The former matrix represent the source re-arrangement commented in the first point above, whereas the latter is a mathematical representation of the scaling factors referred to in the second place. Hypothesis A4 restricts D to have unit-norm diagonal elements. In such a case, the product PD is termed **quasiidentity** matrix [17]. Vectors $\hat{\mathbf{x}}(t)$ and $\mathbf{x}(t)$ related by (4.13) are said to be equivalent in the **waveform-preserving** (**WP**) sense [58–60]. The extraction of the sources is usually carried out by estimating a linear transformation $W \in \mathbb{R}^{q \times p}$ such that

$$\mathbf{s}(t) = W\mathbf{y}(t) \tag{4.14}$$

is waveform-preserving related to the original set of sources $\mathbf{x}(t)$. A matrix W fulfilling this condition is referred to as a **separating matrix**. The **global system**

$$G = WM = (g_{ij})_{q \times q} \tag{4.15}$$

is the matrix which links the original set of sources with the estimated set, that is, the matrix that represents the whole mixing-unmixing system. In the noiseless case, the global matrix exhibits a quasiidentity structure for any valid separating matrix, providing at the separator output a set of signals WP-equivalent to the true sources.

4.2.4   General approach. Identifiability

Most BSS methods turn the sources independence assumption A3 into the separation criterion. A linear transformation W is sought so that the signals at the output of the unmixing system $\mathbf{s}(t)$ are independent. Recall that the independence hypothesis imposes certain conditions on the sources cross-cumulants, given by (4.7). Then, it seems sensible to employ the cumulants to quantitatively measure the degree of independence among the components of $\mathbf{s}(t)$. Specifically, the elements of W must be such that the cross-cumulants of $\mathbf{s}(t)$ vanish. The order up to which these cross-cumulants must be close to zero directly depends on the order of independence assumed for the sources.

The question that naturally arises from these considerations is: "is it sufficient to constrain independence at the separator output to obtain a good estimation of the sources?". The answer is yes, it is. It is shown in [58] that the mixture, i.e. the pair $(M, \mathbf{x}(t))$, is identifiable, and thus recoverable, by using this general procedure if and only if there exists at most one Gaussian signal among the sources. This identifiability condition stems from the Gaussianity property of cumulants, which states that the cumulants of order higher than two of a random variable with Gaussian distribution are all null. This bounds the applicability of cumulants of such orders to the separation of Gaussian signals. The above identifiability condition could also be deduced from the fact that any linear combination of normal random variables is also normally distributed, and hence there is no way of separating such Gaussian signals in an additive model like (4.5). The above is a general condition for the identifiability of instantaneous linear mixtures. Actually, particular methods may have their own specific identifiability conditions, but all of them revolve around this generic result.

Now, depending on the statistical information employed — order of cumulants and degree of statistical information of the sources exploited — two major groups of methods for BSS can be distinguished. The first group is composed of second-order techniques, also known as **principal component analysis (PCA)**. PCA methods seek the removal of second-order dependence from the set of observations, thus providing a set of uncorrelated signals at the unmixing system output. The other main group of BSS methods is referred to as **independent component analysis (ICA)**. They aim to look for an output vector whose components are higher-order independent, the term 'higher' meaning more than second order. PCA procedures are simpler than ICA methods, but they are not always able to preserve the sources waveforms. The ICA overcomes this indeterminacy with the higher-order statistical information of the measurements. The use of this extra information improves the separation quality.

Most BSS methods mainly benefit from the spatial diversity of the problem, that is, the fact that distinct sensors output different mixtures of the source signals, and generally any time structure is ignored. In this connection, the BSS problem reduces to the identification of the source-vector probability distribution from the sample distribution of the observation vector [16]. In some cases, however, the time information can also be exploited, which may offer some advantages. For instance, if the sources have different spectral content, the separation can be achieved by means of the second-order statistics of the observation signals (such as correlation in different time lags). This allows Gaussian sources to be retrieved [4, 58, 60].

4.2.5 Geometrical insights

Let us have a quick look at the BSS problem from a geometrical point of view. In the following, the time index will be dropped for convenience. Each observation sample $\mathbf{y}$ may be represented as a point or vector in a p-dimensional vector space. On the other hand, if $\mathbf{m}_j$, $j = 1, \ldots, q$, denote the column vectors of the mixing matrix, then model (4.5) may alternatively be expressed in the noiseless case as

$$\mathbf{y} = \sum_{j=1}^{q} \mathbf{m}_j x_j. \tag{4.16}$$

Hence, the contribution of source x_j to the measurement vector is given by the projection of its amplitude along the direction of the vector $\mathbf{m}_j$. On this account, the columns of the mixing matrix are also termed **source directions**, **transfer vectors** or **source signatures** [11]. The source directions span the *range* or *column space* of M, also known as the **signal subspace** [54], since it contains all possible measurement vectors generated from the source signals via (4.16). Effectively, the measurement vector $\mathbf{y}$ is built up of the vector sum of the sources projected on the source directions. Component-wise, source x_j is present in the ith measurement in a factor of $\mathbf{m}_{ij}x_j$, which is the projection of $\mathbf{m}_j x_j$ on the ith-axis. Figure 4.4 depicts this notion in the 2-source 3-sensor case. The source waveforms can then be recovered by projecting back the measurement vector onto the source directions. Ideally, we are looking for q vectors $\{\mathbf{w}_i \in \mathbb{R}^p, i = 1, \ldots, q\}$ such that

$$\hat{x}_i = \mathbf{w}_i^{\mathrm{T}}\mathbf{y} = \sum_{j=1}^{p} \mathbf{w}_i^{\mathrm{T}}\mathbf{m}_j x_j \qquad (4.17)$$

is equal to the original source x_i. For this to hold:

$$\mathbf{w}_i^{\mathrm{T}}\mathbf{m}_j = \begin{cases} 0, & i \neq j \\ 1, & i = j. \end{cases} \qquad (4.18)$$

Obviously, the separating matrix $W = [\mathbf{w}_1, \ldots, \mathbf{w}_q]^{\mathrm{T}}$ defined by the last equation is the pseudoinverse of the mixing matrix, which unfortunately is unknown. Hence, as commented at the end of section 4.2.1, the recovery of the sources is intimately related to the identification of the transfer matrix. Observe that a *spatial filtering* process is taking place when performing the separation, since each estimated source is obtained through a particular linear combination of the sensor output:

$$\hat{x}_i = \sum_{j=1}^{p} w_{ij}y_j, \qquad (4.19)$$

in which $W = (w_{ij})_{q \times p}$.

4.2.6 An algebraic standpoint: cumulant tensor diagonalization

The BSS problem may also be approached from the perspective of multilinear algebra. The cumulant array (4.8), at any order, is actually a **tensor** as it fulfils the *multilinearity property*. Given an rth-order array $C_{j_1 \ldots j_r}^x$ defined on a q-dimensional coordinate system, it is said to exhibit the mentioned property when, if vector $\mathbf{x}$ is linearly transformed according to $\mathbf{y} = M\mathbf{x}$, then a similar (multi-)linear relationship holds between the associated arrays, specifically:

$$C_{i_1 \ldots i_r}^y = \sum_{j_1 \ldots j_r} m_{i_1 j_1} \ldots m_{i_r j_r} C_{j_1 \ldots j_r}^x. \qquad (4.20)$$

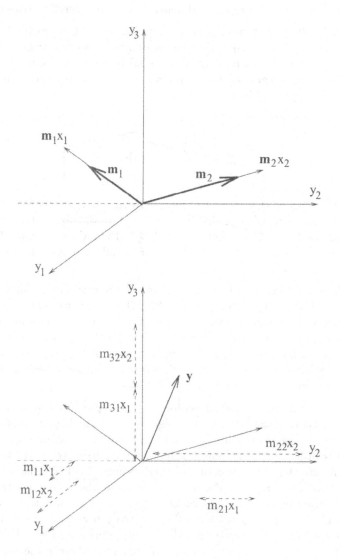

Figure 4.4: Geometrical interpretation of the BSS. In the noiseless two-source three-sensor scenario, the measurement vector **y** can be represented in a three-dimensional space as the vector sum of the source amplitudes, x_1 and x_2 projected along the respective source directions $\mathbf{m}_1$ and $\mathbf{m}_2$.

The cumulant arrays do so [7], and hence can be considered as tensors [43].

Then, in algebraic terms the source extraction and transfer matrix identification is shown to be equivalent to the diagonalization of the observation cumulant tensors [25,58]. Intuitively, this is fairly reasonable, since the source signals being independent, their cumulant tensors possess such a diagonal structure, as explained in section 4.2.2:

$$C^x_{i_1 \ldots i_r} \begin{cases} \neq 0, & i_1 = \cdots = i_r, \\ = 0, & \text{otherwise}, \end{cases} \tag{4.21}$$

where C^x_r is defined in (4.8). Therefore, it seems quite sensible to search for a separating matrix W such that when applied on the observation vector it results in another random vector with diagonal cumulant tensors. Usually it suffices to constrain a diagonal shape for cumulant tensors at one or two different orders only.

In the second-order case, we are aiming at the covariance matrix diagonalization. This is illustrated in figure 4.5. The covariance matrix is always symmetric and, as such, it can be diagonalized by using several procedures. One of them is the eigenvalue decomposition (EVD), although more robust alternative techniques exist, as will be discussed in the next section. The diagonalization of the covariance matrix leads to uncorrelated signals, i.e., only second-order independent.

When $r = 3$, the cumulant tensor may be represented as a cube in a three dimensional space, as shown in figure 4.6. In the higher-order ($r > 2$) case, however, the diagonalization problem becomes more involved. Again, the higher-order cumulant tensor shows a symmetric structure, since the cumulants are symmetric function of their indices [7]. The difficulty lies in the fact that, as opposed to the second-order case where there exists a wide range of algebraic tools to deal with matrices, no such variety is available in the higher-order array domain [25]. On the other hand, the symmetric sensor-output cumulant tensor of order r and p dimensions has $\binom{p+r-1}{r}$ free parameters, introducing just as many constraints, whereas the separating matrix W is composed of qp unknowns. Consequently, only a small subset of symmetric higher-order tensors are linearly diagonalizable [25]. Hence, one is left with two choices: either to try an exact diagonalization in a space of larger dimensions ($q > p$) or to look for an approximate diagonal tensor decomposition in a space of the same dimension ($p = q$) [25]. This issue will be revisited in section 4.6, when dealing with higher-order BSS techniques.

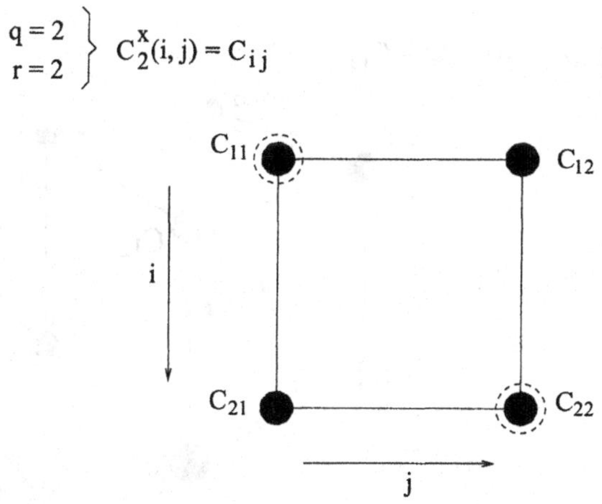

$$\left.\begin{array}{l} q = 2 \\ r = 2 \end{array}\right\} \quad C_2^x(i, j) = C_{ij}$$

Figure 4.5: Second-order cumulant tensor structure. In the second-order case, the cumulant tensor is just the covariance matrix. Making the off-diagonal elements of such matrix zero and preserving non-null diagonal elements, the covariance matrix diagonalization is achieved, thus resulting in uncorrelated components.

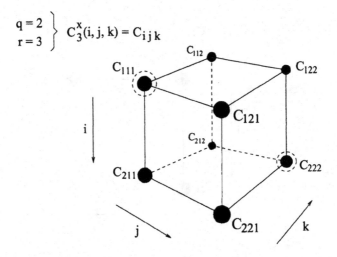

Figure 4.6: Third-order cumulant tensor structure. The elements in the diagonal (surrounded by a dashed circle) are the marginal cumulants. The off-diagonal elements are the cross-cumulants, which must be made zero to achieve independent components. This results in a diagonal cumulant tensor.

4.3 Measuring the separation quality: performance indices

In order to assess the separation quality offered by BSS methods in quantitative terms, some performance indices need to be defined. A variety of such indicators are employed throughout the literature. Herein, we will highlight five of them, which will be useful in evaluating the following aspects of the separation results: a) the similarity between the waveforms of the original sources and the estimated ones, b) the resemblance between the original mixing matrix and the estimated one, c) to what extent the sources interfere in the estimation of each other, d) the degree of statistical independence (at certain order) in the recovered source signals, and e) the computational requirements of a specific method in achieving the separation. Note that in order to obtain the first three measures, the original sources and mixing matrix must be available, while the last two can be used regardless. These performance parameters are all examined next.

4.3.1 Measure of the source-waveform preservation

To measure the resemblance between the original and the reconstructed source waveforms, one can resort to the mean square error of their difference, that is:

$$\varepsilon^2 \triangleq \mathrm{E}\left[(\hat{\mathbf{x}} - \mathbf{x})^2\right]. \tag{4.22}$$

As the sources may be estimated up to a factor PD, the signals in (4.22) should be properly arranged and scaled before computing ε^2. This can be automatically done, for instance, by means of the spatial cross-covariance matrix $\mathrm{cov}[\mathbf{x}, \hat{\mathbf{x}}] = \mathrm{E}\left[\mathbf{x}\hat{\mathbf{x}}^\mathrm{T}\right]$, which indicates in which position of the recovered source vector the estimate of each original source signal appears. When ε^2 is normalized with respect to the source power, it is named **crosstalk** [20, 40, 41]. Assuming that the signals are unit-variance, index ε^2 can be developed into:

$$\varepsilon^2 = \frac{1}{q} \sum_{i=1}^{q} \mathrm{E}\left[(x_i - \hat{x}_i)^2\right] = 2\left(1 - \frac{1}{q} \sum_{i=1}^{q} \mathrm{E}[x_i\hat{x}_i]\right). \tag{4.23}$$

Hence, the better the separation, the closer this parameter is to zero, because in that case, $\hat{\mathbf{x}} \approx \mathbf{x}$ and hence $\mathrm{E}[x_i\hat{x}_i] \approx 1$. On the other hand, if the recovered source waveforms do not look much like the original ones, they are both uncorrelated in statistical terms, and then $E[x_i \hat{x}_i] \approx 0$, so $\varepsilon^2 \approx 2$. In a logarithmic scale, ε^2 can be defined as:

$$\varepsilon^2 \, (\text{dB}) = 10 \log_{10} \varepsilon^2. \qquad (4.24)$$

4.3.2   Quality of the estimated mixing matrix

The following indicator was introduced in [24] to measure the distance between two matrices, modulo a post-multiplicative factor of the form PD. Given two matrices A and B:

$$\epsilon(A, B) \triangleq \sum_i \left| \sum_j |D_{ij}| - 1 \right|^2 + \sum_j \left| \sum_i |D_{ij}| - 1 \right|^2 +$$

$$+ \sum_i \left| \sum_j |D_{ij}|^2 - 1 \right| + \sum_j \left| \sum_i |D_{ij}|^2 - 1 \right| \quad (4.25)$$

with

$$D = \widehat{A}^* \widehat{B}, \qquad \begin{cases} \widehat{A} = A \Delta_A^{-1}, & \Delta_A(k,\, k) = \|A(:,\, k)\| \\ \widehat{B} = B \Delta_B^{-1}, & \Delta_B(k,\, k) = \|B(:,\, k)\|. \end{cases} \qquad (4.26)$$

Since

$$\epsilon(A, B) = 0 \quad \Leftrightarrow \quad B = APD, \qquad (4.27)$$

by using $\epsilon(M, \hat{M})$ the resemblance between the true mixing matrix M and the one estimated by the separation algorithm $\hat{M}$ may be calculated. In a good source extraction, $\epsilon \approx 0$. In dB:

$$\epsilon \, (\text{dB}) = 10 \log_{10} \epsilon. \qquad (4.28)$$

4.3.3   Interference-to-signal ratio

The parameter

$$\varrho_{ij} \triangleq \frac{(WM)_{ij}^2 \, E[x_j^2]}{(WM)_{ii}^2 \, E[x_i^2]}, \qquad i \neq j, \qquad (4.29)$$

quantifies the relative power contribution of the jth source in the estimation of the ith source. It is hence termed **interference-to-noise ratio (ISR)**

[16,18] obtained in rejecting the jth source in the estimate of the ith source by a separating matrix W. For this reason it is also known as **rejection rate** [12,17], and in the terminology of channel equalization as **intersymbol interference (ISI)**. With the unit-variance source convention A4:

$$\varrho_{ij} = \frac{|g_{ij}|^2}{|g_{ii}|^2}, \qquad (4.30)$$

g_{ij} denoting the global matrix coefficients (4.15). Its mean value over the indices ij can be taken as another measure of separation quality:

$$\text{ISR} = \mathop{\text{E}}_{ij}[\varrho_{ij}], \qquad (4.31)$$

and in dB it is defined in a similar way as ε^2 and ϵ above.

4.3.4 Degree of statistical independence

The degree of statistical independence at order r of the components of the random vector $\hat{\mathbf{x}}$ may be measured as the ratio of sum of the squares of their rth-order marginal cumulants to the sum of the squares of all rth-order cumulants:

$$\Upsilon_r \triangleq \frac{\sum_i |\kappa^{\hat{x}}_{i_1 \dots i}|^2}{\sum_{i_1 \dots i_r} |\kappa^{\hat{x}}_{i_1 \dots i_r}|^2}, \qquad \kappa^{\hat{x}}_{i_1 \dots i_r} \triangleq \text{Cum}[\hat{x}_{i_1}, \dots, \hat{x}_{i_r}]. \qquad (4.32)$$

In compliance with section 4.2.6, this performance index is also a measure of how close the rth-order cumulant tensor of the recovered sources is to a diagonal structure. The original sources being independent, in a good separation Υ_r will be very close to one, since the cross-cumulants of independent signals vanish. Sometimes, specially for comparison purposes, it is more useful to consider the 'distance to independence' on a logarithmic scale, that is,

$$1 - \Upsilon_r \,(\text{dB}) = 10 \log_{10}(1 - \Upsilon_r). \qquad (4.33)$$

For the same degree of statistical independence, index Υ_r depends on the relative variance of the components involved. For meaningful measures, it is then convenient to normalize the signals (e.g., to unit variance) before computing this parameter.

4.3.5 Computational complexity

The floating point operations taken by an algorithm to supply a separation solution can be regarded as a good measure of its computational complexity. A floating point operation (or, abbreviated, *flop*) usually corresponds to a multiplication followed by an addition [24].

4.4   A real-life problem: the fetal ECG extraction

Before undertaking the study of methods for BSS, let us first describe an important biomedical problem which will be useful in comparing the performance of the different BSS procedures in the context of a real-world application. This problem is the **fetal ECG extraction**, and it arises when the fetal heart is to be monitored during pregnancy. Usually *non-invasive techniques* are employed, which involve measurements from electrodes attached to different points of the mother's skin. As a result, the recordings pick up a mixture of fetal ECG (FECG) and maternal ECG (MECG) contributions. In addition, due to the low level of the FECG signals other random disturbances become important and must be considered: mains coupling, maternal electromyogram (EMG) interference, thermal noise from the electrodes and other electronic equipment, etc. Hence, the purpose consists in extracting the wanted fetal contributions from the rest of non-desired components that corrupt the cutaneous recordings.

Next section shows that BSS is relevant to this problem. Section 4.4.2 depicts the real ECG data that will allow us to compare the performance of the BSS methods within this context.

4.4.1   BSS model of the FECG extraction problem

Several considerations suggest that the FECG extraction may be formulated as a blind identification problem [46,68]. In fact, this situation may be framed in the field of BSS as follows. Let us start by assuming that the recordings consist of p signals which are obtained by p electrodes attached to different points of the mother's skin. Vector $\mathbf{y}(k)$ contains the kth sample of the skin-electrode signals. Now, bearing in mind the bioelectrical phenomena ruling the cardiac activity and the propagation of heartbeat signals across the body, it can be proven [8,38,51,52] that the maternal ECG signals can be composed as a linear combination of m statistical independent signals ($m = 3$). These independent signals generate the MECG-subspace, or set of all possible ECGs that can be recorded from the mother. Accordingly, it is also claimed [48] that every ECG measured from the fetus can be composed as a linear combination of f signals (usually $f = 2$), which span the FECG-subspace. These vectors that compose the ECG-subspaces are the source signals of the problem and can be written as the vector $\mathbf{x}$ of (4.5) of q components, where $q = m + f$. The linear combination coefficients form the matrix M ($p \times q$) of (4.5) which represents the transfer from the bioelectric sources to the electrodes.

Specifically, its element m_{ij} is the coefficient by which the jth source signal contributes to the ith recording, and is determined by the geometry of the body, the electrode and source positions and the conductivity of the body tissues [61]. If the additive measurement noise is also taken into account, it may be represented by a p-component vector **n** and then we have exactly a problem like (4.5). As a conclusion, the separation of the fetal and maternal ECGs from skin electrodes located on the mother's body can be modelled as a BSS problem: given samples of the p data signals (vector **y**), estimate the source signals (vector **x**) and the coefficients of the linear transformation (matrix M). The estimations of the FECG sources gives FECG contributions to each recording free from MECG and other disturbances. Effectively, if f_1 and f_2 represent the position of the estimated FECG sources, then the fetal heartbeat contribution to each recording is given by:

$$\mathbf{y}_f = \sum_{j=f_1, f_2} \hat{\mathbf{m}}_j \hat{x}_j \qquad (4.34)$$

where $\hat{\mathbf{m}}_j$ stands for the jth column of the estimated transfer matrix and $\hat{x}_i$ the jth obtained source signal. A more detailed explanation on this problem and how it is related to BSS can be found in [1, 2, 46, 68]. The main point is that BSS techniques can be applied to tackle it.

4.4.2 A real 8-channel ECG recording

The experimental data consists of 8 recordings taken from 8 electrodes attached to different points of a pregnant woman's skin. This multi-channel ECG recording is displayed in figure 4.7. The signals have been sampled for 10 seconds at a sampling rate of $f_s = 500$ Hz, and so each signal consists of $T = 5000$ samples. The actual magnitude of the y-axis is not important, but only the relative amplitudes between the recordings. The first five signals have been measured from the abdominal region and even though the fetal ECG is much weaker than the maternal one, it can still be detected. The last three recordings are taken from the mother's thoracic region and, therefore, in them the fetal ECG is not detectable at all. This real 8-channel ECG data will be used as an input matrix to the algorithms which implement the methods that will be described next.

4.5 Methods based on second-order statistics

BSS methods based on SOS are unable to recover the source waveforms if the temporal information is ignored. Despite this fundamental shortcoming,

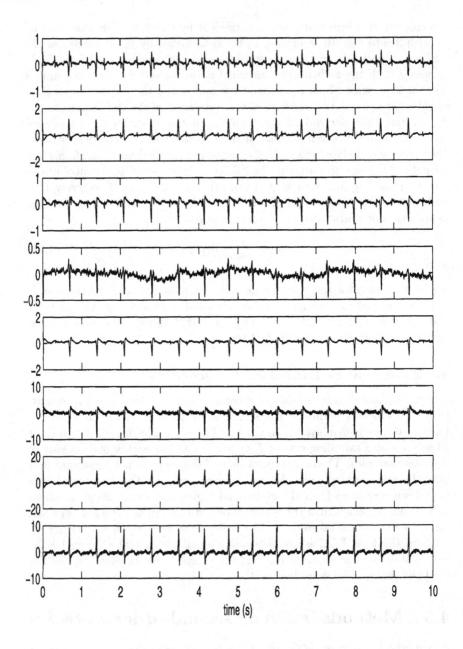

Figure 4.7: 8-channel ECG recording.

it is important to know them, since they are often used by HOS-based techniques as a first step to achieve the separation. The *principal component analysis (PCA)*, as a typical second-order technique, is first summarized in section 4.5.1. Later, section 4.5.2 makes some comments on the lack of its ability in performing waveform-preserving source extraction, and hence is the need for methods based on HOS.

4.5.1 Principal component analysis (PCA)

This is the standard statistical technique based on the decomposition of the covariance matrix of the observed variables [32, 42]. The original aim is to look for a few linear combinations of the observations that may be used to summarize the data, losing in the process as little information as possible. The selected transformation is such that it makes these linear combinations uncorrelated (i.e., independent at the second-order) and with decreasing variance. The resultant signals are regarded as an estimation of the true sources.

Let us assume that the mixing matrix has already been obtained. Then, the source signals may be estimated, for instance, via the maximum-likelihood (ML) principle by maximizing the likelihood of the observations with respect to the source vector. The log-likelihood is given by:

$$L(\mathbf{y}; \mathbf{x}) = \log(p_y(\mathbf{y}; \mathbf{x})) = \log(p_n(\mathbf{y} - M\mathbf{x})), \qquad (4.35)$$

where $p_y(\cdot)$ and $p_n(\cdot)$ denote, respectively, the sensor-output and the noise probability density functions (pdf's). If the noise is Gaussian, the maximizer of (4.35) is readily determined as [54]:

$$\hat{\mathbf{x}}_{\text{ML}} = \left(M^{\text{T}} R_n^{-1} M\right)^{-1} M^{\text{T}} R_n^{-1} \mathbf{y}, \qquad (4.36)$$

where $R_n \triangleq \text{E}[\mathbf{n}\mathbf{n}^{\text{T}}]$ represents the noise covariance matrix. Note that the above expression even handles the case where the noise signals are spatially correlated, provided they are Gaussian and their covariance matrix is available. On the other hand, if its known that the noise signals have the same variance, say σ_N^2, the ML estimator of the sources (4.36) reduces to:

$$\hat{\mathbf{x}}_{\text{ML}} = (M^{\text{T}} M)^{-1} M^{\text{T}} \mathbf{y}, \qquad (4.37)$$

which corresponds to the well-known least-squares (LS) solution [54]:

$$\hat{\mathbf{x}}_{\text{LS}} = \arg \min_{\mathbf{x}} \|\mathbf{y} - M\mathbf{x}\|^2. \qquad (4.38)$$

PCA via EVD

In order to attain explicit PCA estimators of M and $\mathbf{x}$, consider the EVD of the sensor-output covariance matrix, $R_y \triangleq \mathrm{E}[\mathbf{yy}^\mathrm{T}]$:

$$R_y = \Gamma\Lambda\Gamma^\mathrm{T}, \tag{4.39}$$

in which orthogonal matrix Γ contains the *eigenvectors* of R_y and diagonal matrix Λ contains its *eigenvalues* [29]. From the BSS model (4.5), R_y can also be expanded as a function of the mixture elements:

$$R_y = MR_xM^\mathrm{T} + R_n = MM^\mathrm{T} + R_n = MM^\mathrm{T} + \sigma_N^2 I, \tag{4.40}$$

where convention (4.12) has been borne in mind. Combining the previous two equations results in the estimated mixing matrix:

$$\hat{M} = \Gamma\left(\Lambda - \sigma_N^2 I\right)^{\frac{1}{2}}, \tag{4.41}$$

where the rational power is taken element-wise. From (4.37), the estimated source vector becomes:

$$\hat{\mathbf{x}} = \left(\Lambda - \sigma_N^2 I\right)^{-\frac{1}{2}} \Gamma^\mathrm{T}\mathbf{y}. \tag{4.42}$$

Note, however, that with this solution the covariance matrix of the estimated sources is:

$$R_{\hat{x}} \triangleq \mathrm{E}[\hat{\mathbf{x}}\hat{\mathbf{x}}^\mathrm{T}] = \left(\Lambda - \sigma_N^2 I\right)^{-1}\Lambda \neq I. \tag{4.43}$$

Hence, if noise is present ($\sigma_N^2 \neq 0$), additional scaling is necessary to make the recovered sources agree with convention (4.12). In the noiseless case, estimates (4.41) and (4.42) become, respectively:

$$\begin{cases} \hat{M} = \Gamma\Lambda^{\frac{1}{2}} \\ \hat{\mathbf{x}} = \Lambda^{-\frac{1}{2}}\Gamma^\mathrm{T}\mathbf{y}. \end{cases} \tag{4.44}$$

And now $R_{\hat{x}} = I$.

PCA via SVD

More robust, reliable and computationally efficient PCA techniques exist. In particular, one may take advantage of the singular value decomposition (SVD) of the observations, thus avoiding the explicit computation of their covariance matrix. In a batch processing framework, T samples of the sensor signals are stored in the columns of a $p \times T$ matrix Y, so that $Y(:, k) = \mathbf{y}(k)$.

Accordingly, the corresponding samples of the source and the noise signals would be contained in matrices X and N, respectively. The block-processing BSS model is the matrix equivalent of the vector model (4.5):

$$Y = MX + N. \tag{4.45}$$

The observation sample matrix Y can be expanded in its SVD [29]:

$$Y = U\Sigma V^{\mathrm{T}}. \tag{4.46}$$

Matrices $U \in \mathbb{R}^{p \times p}$ and $V \in \mathbb{R}^{T \times p}$ contain the *left* and *right* singular vectors of Y, respectively. Both are orthogonal: $U^{\mathrm{T}}U = UU^{\mathrm{T}} = V^{\mathrm{T}}V = I_p$. Diagonal matrix $\Sigma \in \mathbb{R}^{p \times p}$ contains in its diagonal the *singular values* of sample matrix Y. Then, the observation covariance matrix accepts two different forms, one as a function of the mixing model (4.5) and the other from the previous SVD expansion:

$$R_y = MM^{\mathrm{T}} + R_n = MM^{\mathrm{T}} + \sigma_N^2 I \tag{4.47}$$

$$R_y \approx \frac{1}{T} YY^{\mathrm{T}} = \frac{1}{T} U\Sigma^2 U^{\mathrm{T}}. \tag{4.48}$$

Combining (4.47) and (4.48), one can easily obtain an estimate of the mixing matrix:

$$\hat{M} = \frac{1}{\sqrt{T}} U \left(\Sigma^2 - T\sigma_N^2 I \right)^{\frac{1}{2}}. \tag{4.49}$$

If the noise is *temporally white* up to the second-order (uncorrelated at different time instants), this expression can be substituted into (4.37), yielding the source components:

$$\hat{X} = \sqrt{T} \left(\Sigma^2 - T\sigma_N^2 I \right)^{-\frac{1}{2}} U^{\mathrm{T}} Y = \sqrt{T} \left(\Sigma^2 - T\sigma_N^2 I \right)^{-\frac{1}{2}} \Sigma V^{\mathrm{T}}. \tag{4.50}$$

Again, with this solution,

$$R_{\hat{x}} \approx \frac{1}{T} \hat{X}\hat{X}^{\mathrm{T}} = \left(\Sigma^2 - T\sigma_N^2 I \right)^{-1} \Sigma^2 \neq I, \tag{4.51}$$

and so further scaling may be necessary in order to preserve the identity structure of the source covariance matrix. That is, the signals extracted are second-order spatially white (mutually uncorrelated), due to the orthogonality of the right singular vectors of Y, but not necessarily equal-power. In the noiseless case, estimates (4.49) and (4.50) become, respectively:

$$\begin{cases} \hat{M} = \frac{1}{\sqrt{T}} U\Sigma \\ \hat{X} = \sqrt{T} V^{\mathrm{T}}, \end{cases} \tag{4.52}$$

where now $R_{\hat{x}} = I$. Observe that the EVD and SVD solution are, in principle, equivalent. The covariance matrix factorizations (4.39) and (4.48) being equal, the relationships

$$\left\{ \begin{array}{l} \frac{1}{\sqrt{T}} U = \Gamma \\ \Sigma^2 = \Lambda, \end{array} \right. \tag{4.53}$$

hold, and then (4.41) and (4.49) are ideally the same estimates. The difference between the two procedures lies in the fact that in many cases the covariance matrix of the observations is ill-conditioned. In such situations, the SVD-based procedure proves more reliable, as it avoids the explicit computation of the sensor-output covariance matrix.

Both EVD solutions (4.41)–(4.42) and SVD solutions (4.49)–(4.50), albeit elegant, lack real feasibility. Effectively, the noise variance σ_N^2 is unknown in a genuine BSS situation. If the number of sensors is greater than the number of sources, i.e., $p > q$, σ_N^2 can be estimated from the smaller singular values of Y, or from the smallest eigenvalues of its covariance matrix. In the likely case in which these smallest singular/eigen values are not equal, their average may be taken. However, in reality the number of sources is unknown and, even if it is not, experiments show that in many cases the singular values of the noise cannot be distinguished from those of the information or desired signals [1].

In conclusion, when nothing is known a priori about the noise, its effects have to be neglected and one has to assume that the **noiseless BSS model**

$$\mathbf{y} = M\mathbf{x} \tag{4.54}$$

is being treated, where the PCA solutions are given by (4.44) and (4.52). Yet further simplifications are still possible. Specifically, each of the p observed signals comes from a linear combination of q statistically independent source signals, which in particular are linearly independent (it is readily seen that uncorrelation implies linear independence). This means that the rank of the sample matrix Y is q, which can be obtained as the number of its singular values significantly greater than zero. As a result, Y accepts a reduced SVD expansion in the form [29]:

$$Y = U_q \Sigma_q V_q^T \tag{4.55}$$

where $U_q = U(:, 1:q)$, $\Sigma_q = \Sigma(1:q, 1:q)$ and $V_q = V(:, 1:q)$. Therefore the mixing matrix M and the source sample matrix X and can be estimated as:

$$\left\{ \begin{array}{l} \hat{M}_{\text{PCA}} = \frac{1}{\sqrt{T}} U_q \Sigma_q \\ \hat{X}_{\text{PCA}} = \sqrt{T} V_q^T, \end{array} \right. \tag{4.56}$$

and analogously for the EVD solutions (4.44).

One final comment about classical statistical nomenclature is in order. Strictly speaking, when the observations follow a noisy model like (4.5)–(4.45), one is actually performing a **Factor Analysis (FA)** on the observed variables to obtain estimates of the mixing matrix and the source signals. Only when the noise is neglected FA and PCA become identical [32, 42].

4.5.2 Justifying HOS: indeterminacy of second-order analysis

In their very nature, 2nd-order techniques, such as the PCA seen earlier, only resort to the 2nd-order statistical information of the given dataset. Hence the estimation of the sources carried out by such techniques results, at most, in signals with diagonal covariance matrix (spatial uncorrelation or 2nd-order spatial whiteness). Assuming the constraint (4.12), the estimated-source covariance matrix satisfies:

$$R_s \triangleq \mathrm{E}\big[\mathbf{s}\mathbf{s}^{\mathrm{T}}\big] = I_q, \qquad (4.57)$$

where $\mathbf{s} = W\mathbf{y}$ represents the reconstructed sources. However, suppose that an orthogonal transformation $\tilde{Q}$ is now applied to the outputs $\mathbf{s}$. In such a case, the new set of signals, say $\mathbf{s}' = \tilde{Q}\mathbf{s}$ would also satisfy the constraint, since

$$R_{s'} = \tilde{Q}R_s\tilde{Q}^{\mathrm{T}} = I_q, \qquad (4.58)$$

due to the orthogonality of the matrix $\tilde{Q}$. Consequently the sets of estimated sources contained in $\mathbf{s}$ and $\mathbf{s}'$ are, as far as the fulfilment of the uncorrelation hypothesis is concerned, equivalent. Nevertheless, it is clear that their waveforms are not necessarily the same, and neither is guaranteed that any of these solutions is equivalent to the original sources in the WP sense. This result is due to the fact that there remains an indeterminacy in the form of an orthogonal matrix ($\tilde{Q}$ above), which 2nd-order analysis is unable to identify. In short, by means of 2nd-order analysis the mixing matrix M can only be identified up to post-multiplication by an orthogonal matrix or, equivalently, the sources $\mathbf{x}$ can only be identified up to pre-multiplication by the same orthogonal matrix transpose. As a matter of fact, considering the SVD of the mixing matrix it is simple to realize that by means of second-order techniques only the left singular vectors and the singular values of M are identified, but its right singular vectors cannot be revealed. In turn, this outcome suggests that one could factorize the mixing matrix as:

$$M = BQ \qquad (4.59)$$

where B is a $p \times q$ non-singular matrix and Q a $q \times q$ orthogonal matrix. The former comprises the left singular vectors and the singular values of M, whereas the latter represents the right singular matrix of M. This matrix factorization is called *polar decomposition* [29]. Now, the observation covariance matrix may be expressed as:

$$R_y = M R_x M^{\mathrm{T}} + R_n = B B^{\mathrm{T}} + R_n. \tag{4.60}$$

Looking back at (4.47) it turns out the solutions given by the PCA are actually the matrix B and a set of **whitened observations**, $\mathbf{z}$, rather than the actual sources and mixing matrix:

$$\begin{cases} B = \hat{M}_{\mathrm{PCA}} \\ \mathbf{z} = \hat{\mathbf{x}}_{\mathrm{PCA}}. \end{cases} \tag{4.61}$$

That is, second-order analysis is only able to identify, at most, the singular values and left singular vectors of the mixing structure, contained in matrix B, and a set of uncorrelated signals, in vector $\mathbf{z}$. PCA identifies the source signal subspace, spanned by the columns of M. Note, though, that in order to recover the source waveforms undistorted we need the exact source directions themselves (i.e., a very specific base of such subspace) so that we can project the measurement amplitude vector back onto them, as seen in section 4.2.5.

The signals in $\mathbf{z}$ can be obtained from the sensor output as:

$$\mathbf{z} = B^* \mathbf{y}, \tag{4.62}$$

the symbol $*$ denoting the Moore-Penrose pseudoinverse [6,29]. Matrix B^* is called **whitening matrix**, since its application on the sensor output results in a set of spatially uncorrelated (or spatially *white* at 2nd-order) components. Accordingly, this 2nd-order processing is often referred to as **whitening**. In the noiseless case the whitened sensor-output $\mathbf{z}$ and the source signals $\mathbf{x}$ are then related via the expression:

$$\mathbf{z} = Q\mathbf{x}, \tag{4.63}$$

which points out that the orthogonal matrix Q, representing the right singular vectors of the transfer matrix, remains to be unveiled. The estimation of Q leads to a complete identification of the mixing structure (the columns of M) and the source waveforms, up to the WP equivalence.

In order to disclose this transformation, more constraints are needed, and these come from exploiting the source statistical independence assumption. That is precisely the purpose of higher-order methods. After a first estimation of the sources via a 2nd-order procedure, the matrix Q is sought

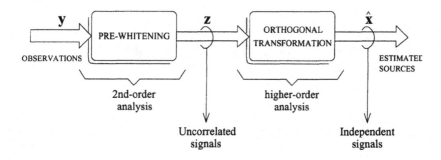

Figure 4.8: The two-stage approach to signal separation.

by constraining higher-order independence at the outputs. This process involves, explicitly or implicitly, resorting to the higher-order information (the higher-order cumulants) of the signals. The new search will supply a new set of signals which, under the general identifiability conditions established in section 4.2.4, will be WP equivalent to the wanted sources. In conclusion, the indeterminacy vanishes with the use of HOS techniques. Figure 4.8 sketches this two-step approach to BSS.

The indeterminacy of SOS-based methods may be looked at from another perspective. By virtue of formulae (4.41) and (4.49) (or their noiseless-model counterparts in (4.44) and (4.52)), the mixing matrix estimated by second-order methods has always orthogonal columns (scaled versions of the left singular vectors of M contained in matrix B above). In other words, second-order methods are only able to estimate mixing matrix with such a structure, which is certainly not a valid solution if the true mixing matrix does not exhibit that feature. Figure 4.9 illustrates this issue in a noiseless two-source two-sensor separation scenario. Each cross represents a sample of the sensor output, for which its abscissa value corresponds to the amplitude of the first sensor and its ordinate value the amplitude of the second sensor: $(y_1(k), y_2(k))$. This type of plot is called *scatter diagram* or *scatter plot*, and in this case it is the two-dimensional equivalent of figure 4.4 but for a number of *snapshots* (time samples). The mixing-matrix columns are chosen non-orthogonal, but the source directions estimated by the PCA (via SVD) are orthogonal, thus not achieving a good separating solution. By contrast, HOS-based techniques can identify a mixing matrix of any structure, and so they do not suffer indeterminacy difficulties, as we will see in section 4.6.

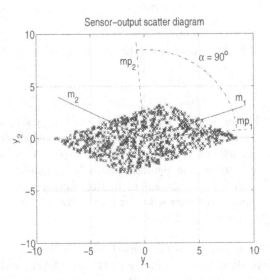

Figure 4.9: Indeterminacy of the second-order methods. The plot displays a 1000-sample scatter diagram of a mixture of two uniformly distributed sources. Vectors m_1 and m_2 represent the true source directions, chosen to be non-orthogonal. Vectors m_{p_1} and m_{p_2} show the source directions obtained by the PCA. They are orthogonal.

4.5.3 Examples

Synthetic signals.

Figure 4.10 plots three source signals composed of 1000 samples: a sinusoid, a binary sequence (pseudorandom with equiprobable symbols) and a triangular signal. A regular mixing matrix is randomly chosen as:

$$M = \begin{bmatrix} -0.4 & 0.1 & -0.4 \\ -0.2 & -0.4 & 0.9 \\ -0.2 & 0.5 & 0.6 \end{bmatrix}. \tag{4.64}$$

The product of the mixing matrix and the source signals yields the set of measurements displayed in figure 4.11. From this set of sensor-output samples, the PCA method (implemented with the SVD version described in section 4.5.1) obtains the set of sources represented in figure 4.12. As seen in this plot, the recovered source waveforms do not match the original ones, but there remain significant contributions of the other sources in the estimate of each one of them. This outcome was expected from the comments made in the previous sections. For this separation, the particular values of the quality indicators introduced in section 4.3 appear in the first row of table 4.1 on page 245. The figures in parentheses are expressed in dB, while the rest are in a linear scale.

Fetal ECG extraction.

The source signals retrieved by the PCA method from the ECG recordings of figure 4.7 can be seen in figure 4.13. The PCA method reveals only two clear MECG signals (1st and 2nd waveforms in fig. 4.13), while the other one (4th signal in fig. 4.13) is heavily contaminated by noise. As far as the FECG sources are concerned (5th and 7th signals in fig. 4.13), the first one presents an undesirable low-frequency fluctuation, which is possibly related to the mother's respiration. It is also observed that the obtained noise sources (3rd, 6th and 8th waveforms in fig. 4.13) exhibit a series of peaks that correspond to the MECG and FECG signals, and so they cannot be considered as pure noise. Note that one out of the last three recovered sources in figure 4.13 corresponds to a FECG source, and therefore its singular value could not be taken to estimate the noise singular values. This experimental result agrees with the comments made in the previous section. As a quantitative measure of the separation quality achieved, the values for some of the performance indexes defined in section 4.3 are calculated as shown in the first column of table 4.2 in page 246.

In [61] a thorough study of the PCA indeterminacy is carried out in connection to this biomedical application. It is seen how the indeterminacy problems can be palliated by making an appropriate choice of the electrode

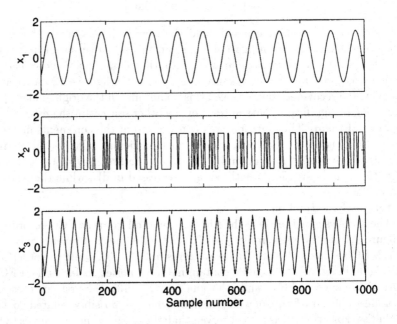

Figure 4.10: 1000 samples of a three-source realization: a sinusoid, a binary sequence and a triangular waveform.

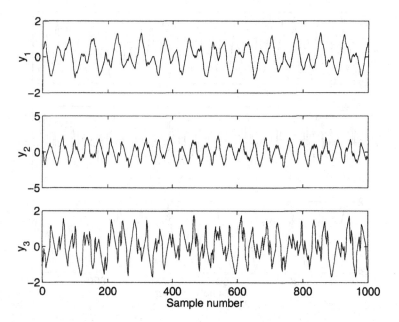

Figure 4.11: Instantaneous linear mixture of the sources shown in figure 4.10.

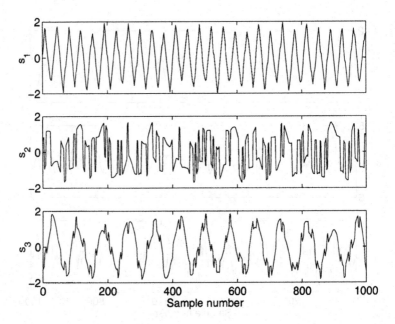

Figure 4.12: Source signals estimated by the PCA (via SVD) method from the mixtures of figure 4.11.

positions. Specifically, the performance of the PCA is shown to highly depend on the chosen electrode placement. This dependency is removed with the use of HOS procedures.

4.6 Methods based on higher-order statistics

In the previous section, the need for HOS-based procedures has been justified if a complete WP source extraction and mixing matrix identification is to be obtained when the temporal information is disregarded. BSS methods relying on HOS are commonly gathered under the name of *independent component analysis (ICA)*. The goal of an ICA technique is to maximize the degree of statistical independence (up to a given order) among the output components. The various methods differ in the way in which this objective is sought. We will mainly focus on batch-processing methods, such as the *higher-order EVD*, the *higher-order SVD* and the *extended ML*, although adaptive separation is also considered in a section devoted to *equivariant* adaptive methods. The adaptive version of some batch procedures is not difficult to derive, though.

4.6.1 Higher-order eigenvalue decomposition (HOEVD)

This method was originally established in [24], where the concept of ICA was first given a rigorous mathematical definition. The method implicitly relies on the independence criterion based on cumulants: a set of random variables are statistically independent up to order r if and only if all their sth-order cross-cumulants are null, for $s \leqslant r$. The sources estimated through this procedure are realizations of a random vector whose components are the "most independent possible", in the sense of the maximization of a given contrast function. A source vector obtained by means of this criterion is said to come from the ICA of the observation vector.

Contrast functions

Let ϵ_n the space of real random variables of dimension n admitting a density. A **contrast function** is a mapping Ψ from the set of densities $\{p_z, \ \mathbf{z} \in \epsilon_n\}$ to $\mathbb{R}$ satisfying [24, 25]:

1. $\Psi(p_z)$ does not change if the components z_i are permuted: $\Psi(p_{Pz}) = \Psi(p_z)$, for any permutation matrix P.

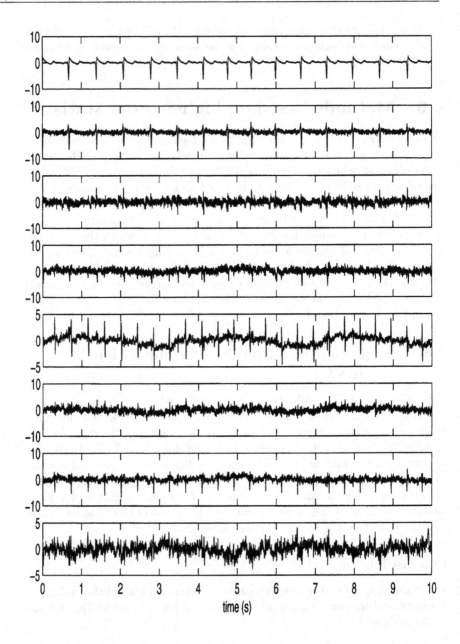

Figure 4.13: Source signals retrieved by the PCA (via SVD) method from the electrode recordings shown in figure 4.7.

2. $\Psi(p_z)$ is invariant under scale change: $\Psi(p_{Dz}) = \Psi(p_z)$, for any invertible diagonal matrix D.

3. $\Psi(p_z)$ decreases by linear combination: if $\mathbf{x}$ has independent components, then $\Psi(p_{Ax}) \leqslant \Psi(p_x)$, for all invertible matrix A.

In addition, a contrast function is said to be *discriminating* over a set ϵ if the equality $\Psi(p_{Ax}) = \Psi(p_x)$ holds only when $A = PD$, $\mathbf{x}$ being a random vector of ϵ with independent components.

A random vector is composed of independent components if and only if their joint pdf (jpdf) can be expressed as the product of their marginal pdf's. Therefore, a natural measure of independence is the *Kullback-Leibler divergence* between the pdf of the separator output and the pdf of the output if it was made up of independent components [27]:

$$\delta(p_s, p_{s'}) = \int p_s(\mathbf{u}) \log \frac{p_s(\mathbf{u})}{\prod\limits_{i=1}^{q} p_{s_i}(\mathbf{u})} \, d\mathbf{u}, \qquad \mathbf{u} \in \mathbb{R}^q, \qquad (4.65)$$

where $p_s(\mathbf{s})$ is the pdf of the random vector $\mathbf{s}$ and $p_{s_i}(s_i)$ the marginal pdf of its ith component. This turns out to be the average mutual information of the estimated sources:

$$\Psi(p_s) = -I(p_s) = -\delta(p_s, p_{s'}). \qquad (4.66)$$

It is shown that Ψ defined as above is a contrast, and is discriminating over the set of random variables having at most one Gaussian component [24].

As $\Psi(p_s)$ involves, in practice, an unknown density function, p_s is approximated by its Edgeworth expansion [43,57] as a function of the cumulants of $\mathbf{s}$. After some simplifications, the contrast function becomes:

$$\psi_r = \sum_{i=1}^{q} (C^s_{\underbrace{ii\ldots i}_{r}})^2 \qquad (4.67)$$

in which $C^s_{\underbrace{ii\ldots i}_{r}}$ represents the rth-order marginal standardized cumulant

of the variable s_i. The term "standardized" (or, equivalently, "normalized") means that the initial set of observed variables has already been transformed into another set with identity covariance matrix, i.e., a diagonalization and normalization of the observation covariance matrix has been carried out previously (2nd-order processing). It may seem surprising that a criterion only based on marginal cumulants works correctly. However, taking into account that

$$\Omega = \sum_{i_1 i_2 \ldots i_r} (C^s_{i_1 \ldots i_r})^2, \qquad C^s_{i_1 \ldots i_r} = \mathrm{Cum}[s_{i_1}, s_{i_2}, \ldots s_{i_r}], \qquad (4.68)$$

is invariant under linear and invertible transformations preserving standardization (orthogonal transformations) then the maximization of ψ is equivalent to the minimization of $\Omega - \psi$, i.e., of the sum of the squares of all rth-order cross-cumulants. These cumulants are precisely the measure of statistical dependence at that order. In short, the maximization of ψ leads to the maximum statistical independence at order r, as claimed at the beginning. Actually, it is shown [24, 25] that (4.67) is a contrast, and is discriminating over the subset of ϵ_n of random vectors with finite moments up to order r and with at most one component with null marginal cumulant of order r.

Pairwise versus mutual independence

Having performed the pre-whitening stage, an orthogonal transformation $\tilde{Q}$ is sought to maximize (4.67). When a contrast of polynomial form is utilized, it suffices to consider only pairwise independence. At first glance, **pairwise independence** could be regarded as a weaker criterion than **mutual independence**, but the following theorem guarantees that both notions are synonymous in our problem [24]:

Theorem 1
Let $\mathbf{x}$ be a vector with non-deterministic independent components, of which at most one is Gaussian. Let G be an orthogonal $q \times q$ matrix, and $\mathbf{s}$ the vector $\mathbf{s} = G\mathbf{x}$. Then, the following three properties are equivalent:

(i) The components s_i are pairwise independent.

(ii) The components s_i are mutually independent.

(iii) $G = PD$, P permutation, D diagonal.

Thus a pairwise processing will provide the desired results while reducing the computational complexity of the method. Although this criterion is derived after simplifications, has the major advantage that it leads to an algorithm with polynomial execution time. For two variables and considering 4th-order statistical independence, the criterion (4.67) becomes:

$$\psi_4(\tilde{Q}; \mathbf{s}) = (C^{s'}_{iiii})^2 + (C^{s'}_{jjjj})^2, \qquad (4.69)$$

where we denote $\mathbf{s} = [s_i, s_j]^T$ and $\mathbf{s}' = [s_i', s_j']^T$ the output variables before and after the maximization, related through $\mathbf{s}' = \tilde{Q}\mathbf{s}$. The 4th-order marginal cumulant is also termed kurtosis (see chapter 1), and hence the criterion tries to maximize the sum of square kurtosis of each output signal pair. Transformation $\tilde{Q}$ is the Givens rotation matrix (counter-rotation of angle θ):

$$\tilde{Q} = \begin{bmatrix} \cos\theta & \sin\theta \\ -\sin\theta & \cos\theta \end{bmatrix} = \frac{1}{\sqrt{1+\alpha^2}} \begin{bmatrix} 1 & \alpha \\ -\alpha & 1 \end{bmatrix}, \qquad \alpha = \text{tg}\,\theta \qquad (4.70)$$

and the angle θ is obtained from the maximization of (4.69). Expanding the expression (4.69) as a function of α, one can arrive at:

$$\psi_4(\zeta) = (\zeta^2 + 4)^{-2} \sum_{k=0}^{4} b_k\,\zeta^k \qquad (4.71)$$

where $\zeta = \alpha - 1/\alpha$ and the coefficients b_k are given from non-linear combinations of the pairwise 4th-order cumulants of s_i and s_j. In order to maximize (4.71), its derivative must be found and set equal to zero:

$$\omega(\zeta) = \sum_{k=0}^{4} c_k\,\zeta^k = 0 \qquad (4.72)$$

in which the coefficients c_k can also be written as a function of the 4th-order cumulants of the same pair of signals. The exact expressions for b_k and c_k can be found in the appendices of [24]. In particular, in the noiseless case this last polynomial degenerates into another one of degree 2 only. Then, extracting the roots of (4.72) and substituting them in (4.71) to verify which one yields the maximum value, the optimum rotation angle is obtained.

Extension to general BSS set-up

To extend this to more than two signals, a first estimation of the transfer matrix and the source signals are found from 2nd-order analysis (PCA):

$$\begin{cases} \hat{M}_0 = \hat{M}_{\text{PCA}} = B \\ \mathbf{s}_0 = \hat{\mathbf{x}}_{\text{PCA}} = \mathbf{z}. \end{cases} \qquad (4.73)$$

Then the above procedure for the maximization of the pairwise 4th-order independence is applied to all the $q(q-1)/2$ pairs of estimated sources. After every calculation of the optimum rotation angle, the $\hat{M}_l$ and $\mathbf{s}_l$ are updated as $\hat{M}_l := \hat{M}_{l-1}\tilde{Q}_l^T$ and $\mathbf{s}_l := \tilde{Q}_l\mathbf{s}_{l-1}$, where $\tilde{Q}_l$ is the rotation matrix found at iteration l. Matrix $\tilde{Q}_l \in \mathbb{R}^{q \times q}$ is similar to (4.70), but it is actually a rotation in the plane (s_i, s_j) when this pair of signals is processed. All this sweep over the output signals pairs is repeated until the convergence of the algorithm, which is estimated to take about $L = 1 + \sqrt{q}$ sweeps. Final estimates of the mixing matrix and the source signals are obtained as:

$$\begin{cases} \hat{M} = \hat{M}_L = B\tilde{Q}_1^{\mathrm{T}}\tilde{Q}_2^{\mathrm{T}}\ldots\tilde{Q}_L^{\mathrm{T}} \\ \hat{\mathbf{x}} = \mathbf{s}_L = \tilde{Q}_L\ldots\tilde{Q}_2\tilde{Q}_1\mathbf{z}. \end{cases} \tag{4.74}$$

Note that matrix Q is estimated as the product $\tilde{Q}_1^{\mathrm{T}}\tilde{Q}_2^{\mathrm{T}}\ldots\tilde{Q}_L^{\mathrm{T}}$.

Links and remarks

The procedure outlined above is called HOEVD because it may be seen as a higher-order extension of the Jacobi method for the computation of the EVD of a Hermitian matrix (second-order symmetric tensor) A [39]. The Jacobi method tries to minimize the sum of squares of off-diagonal elements of the transformed matrix $A' = \tilde{Q}A\tilde{Q}^{\mathrm{T}}$ [29]:

$$\tilde{Q}_{\mathrm{opt}} = \arg\min_{\tilde{Q}} \sum_{i \neq j} |a'_{ij}|^2. \tag{4.75}$$

Since the Frobenius norm $\mathrm{trace}(A'A'^{\mathrm{T}}) = \sum_{ij}|a'_{ij}|^2$ of the transformed matrix remains constant, the minimization in (4.75) is tantamount to the maximization of the sum of squared diagonal elements:

$$\tilde{Q}_{\mathrm{opt}} = \arg\max_{\tilde{Q}} \psi_2(\tilde{Q}; A), \qquad \psi_2 = \sum_i |a'_{ii}|^2, \tag{4.76}$$

which is the second-order counterpart of criterion (4.67). In particular, (4.67) reduces to (4.76) when A is the covariance matrix of the observations. In this respect, equation (4.76) could be considered as the 'contrast function' for the computation of the PCA via EVD. With contrast (4.69) it is the sensor-output fourth-order cumulant tensor which is diagonalized, thus minimizing the statistical dependence of the output signals at that order.

Interestingly enough, maximization criterion (4.69) was also arrived at in [28] (see also [30]; and refer to [31] for an adaptive version) from a rather disparate approach. Specifically, the ML criterion was adopted in the latter references, and after certain simplifications the same cost function was obtained.

A last question remains to be answered: since the higher the cumulant order, the higher the estimation error for the same sample size, why not use third-order cumulants instead of fourth-order ones in the estimation of Q? The justification of this apparent contradiction stems from the fact that most distributions encountered in practice are symmetrical or near symmetrical, and hence their third-order cumulants are very close to zero. Consequently, the associated cumulant tensors are ill-conditioned, hindering the source identification if such order is used.

Examples

Synthetic signals.

From the mixture of figure 4.11, the ICA-HOEVD method estimated the source signals plotted in figure 4.14. Note the considerable improvement relative to the PCA. The performance indices for this separation is in the second row of table 4.1 in page 245. The quantitative performance measures corroborate the visual comparison: the recovered sources are more independent, more similar to the original ones and the estimated mixing matrix is also more alike the original mixing matrix than for the PCA. This is obtained at the expense of more computations.

It is interesting to observe, also, that the independence measure Υ_4 is larger for the sources estimated by the ICA-HOEVD than for the original sources. For the latter: $\Upsilon_4 = 0.9749$, $1 - \Upsilon_4 = -16.00$ dB. This result may be accounted for by resorting to the concepts of population and sample [57]. When we talk about 'source independence', we are actually referring to independence among their respective populations. However, when drawing particular samples from those populations in order to create specific source realizations, the sample measures of independence (such as the parameter Υ_4) no longer manifest total independence, but are subject to variations which are related to the particular samples themselves (sampling errors). Effectively, the cost function used by the ICA-HOEVD (formula (4.69)) is built up of the cumulants of the separator output, which for finite sample-size are estimated with an unavoidable sampling error [57]. By construction of the algorithm, the ICA-HOEVD tries to maximize that measure of independence among the source components. And it occurs that a transformation may be found for which the sample independence measurement for the transformed signal samples is larger than for the original ones. This simply cannot happen when dealing with populations, since in such a case any independence measure would already be maximal for the original sources. Asymptotically, i.e., as the sample size tends to infinity, by consistency of cumulant sample estimates the sample results would coincide with the population (or 'ideal') results.

Fetal ECG extraction.

Figure 4.15 displays the source signals obtained by the ICA-HOEVD

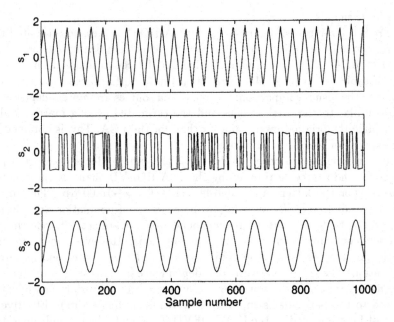

Figure 4.14: Source signals reconstructed by the ICA-HOEVD method from the mixtures of figure 4.11.

method from the skin electrode recordings shown in figure 4.7. The difficulties encountered in the PCA seem to be overcome. In the first place, the noise sources (3rd, 6th and 8th waveforms in fig. 4.15) do not have any contribution from the ECG subspaces, and therefore they can be regarded as pure noise. Secondly, the third MECG signal (4th waveform in fig. 4.15) appears much neater. The same occurs with the 1st MECG source, which is less noisy than its equivalent 2nd MECG source found by the PCA. Hence, the MECG subspace is now clearly revealed. Finally, the fluctuation that was present at the first of the FECG source signals found with the PCA method, now vanishes and forms a separate noise signal, that of the 6th recovered source in fig. 4.15. Moreover, the other FECG signal (5th waveform in fig. 4.15) is slightly less noisy than its equivalent found with the PCA (7th signal in fig. 4.13). The parameters shown in the second columns of table 4.2 in page 246 validate in quantitative terms the comparison made by visual inspection.

4.6.2 Higher-order singular value decomposition (HOSVD)

The HOSVD method for BSS was first presented in [39]. It resorts to a number of concepts about higher-order arrays as well as to a generalization of the matrix SVD. Both topics can be found in the mentioned reference but are also reproduced here for the sake of completeness. These multilinear algebraic tools are then applied to the BSS problem, deriving the HOSVD method. Later in this section, the performance of this method is demonstrated through some examples.

Higher-order arrays and the HOSVD

Higher-order arrays.

An Nth-order array Φ with complex elements is defined as a multidimensional matrix of dimensions $I_1 \times I_2 \times \ldots I_N$: $\Phi \in \mathbb{C}^{I_1 \times I_2 \times \ldots I_N}$. Some operations and properties on higher-order arrays are set out below [39].

SCALAR PRODUCT. The scalar product $\langle \Phi, \Psi \rangle$ of two arrays Φ, $\Psi \in \mathbb{C}^{I_1 \times I_2 \times \ldots I_N}$ is defined as:

$$\langle \Phi, \Psi \rangle \triangleq \sum_{i_1} \sum_{i_2} \cdots \sum_{i_N} \Psi^*_{i_1 i_2 \ldots i_N} \Phi_{i_1 i_2 \ldots i_N}$$

where the symbol $*$ denotes the complex conjugation.

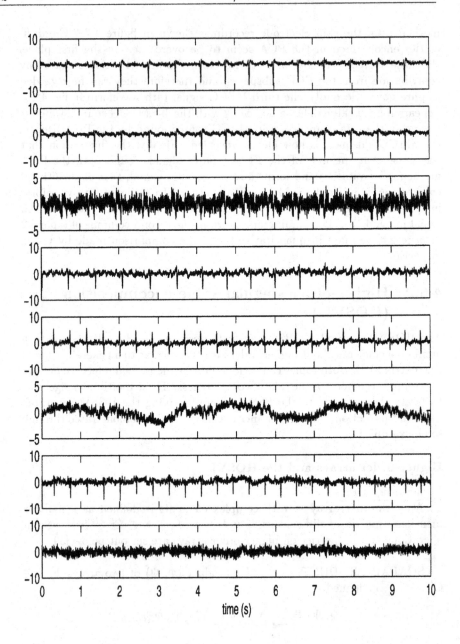

Figure 4.15: Source signals retrieved by the ICA-HOEVD method from the electrode recordings shown in figure 4.7.

ORTHOGONALITY. If $\langle \Phi, \Psi \rangle = 0$, the arrays Φ and Ψ are said to be mutually orthogonal.

FROBENIUS-NORM of an array Φ: $\|\Phi\| \triangleq \sqrt{\langle \Phi, \Phi \rangle}$.

MULTIPLICATION OF A HIGHER-ORDER ARRAY BY A MATRIX. The n-mode product of the Nth-order array $\Phi \in \mathbb{C}^{I_1 \times I_2 \times \dots I_N}$ by the matrix $U \in \mathbb{C}^{J_n \times I_n}$, denoted by $\Phi \times_n U$, is an $(I_1 \times I_2 \times \dots J_n \times \dots I_N)$-array whose entries are given by:

$$(\Phi \times_n U)_{i_1 i_2 \dots j_n \dots i_N} \triangleq \sum_{i_n} \Phi_{i_1 i_2 \dots i_n \dots i_N} \, U_{j_n i_n}.$$

Corollary 1. Given the array $\Phi \in \mathbb{C}^{I_1 \times I_2 \times \dots I_N}$ and the matrices $F \in \mathbb{C}^{J_n \times I_n}$, $G \in \mathbb{C}^{J_m \times I_m}$, it follows that

$$(\Phi \times_n F) \times_m G = (\Phi \times_m G) \times_n F = \Phi \times_n F \times_m G.$$

Corollary 2. Given the array $\Phi \in \mathbb{C}^{I_1 \times I_2 \times \dots I_N}$ and the matrices $F \in \mathbb{C}^{J_n \times I_n}$, $G \in \mathbb{C}^{K_n \times J_n}$, it follows that

$$(\Phi \times_n F) \times_n G = \Phi \times_n (GF).$$

As an example, for the real matrices $S \in \mathbb{C}^{I_1 \times I_2}$, $U \in \mathbb{C}^{J_1 \times I_1}$ and $V \in \mathbb{C}^{J_2 \times I_2}$:

$$USV^{\mathrm{T}} \equiv S \times_1 U \times_2 V. \tag{4.77}$$

Finally, observe that a higher-order tensor is a particular higher-order array which behaves in a specific manner under coordinate transformation of the underlying vector space [43], i.e., which exhibits the multilinearity property (4.20). Using this higher-order array notation, such a property can be written as:

$$C_r^y = C_r^x \times_1 M \times \dots \times_r M. \tag{4.78}$$

The higher-order singular value decomposition.

Every complex $(I_1 \times I_2 \times \dots I_N)$-array Φ can be written as the product [39]

$$\Phi = \Sigma \times_1 U^{(1)} \times_2 U^{(2)} \dots \times_N U^{(N)}$$

where:

- $U^{(n)} = \left[u_1^{(n)}, u_2^{(n)}, \dots, u_{I_n}^{(n)} \right]$ is a complex orthogonal $(I_n \times I_n)$-matrix

- the core array Σ is a complex $(I_1 \times I_2 \times \ldots I_N)$-array whose subarrays $\Sigma_{i_n=\alpha}$ obtained by fixing the nth index to α have the properties of:

  - all-orthogonality: two subarrays $\Sigma_{i_n=\alpha}$ and $\Sigma_{i_n=\beta}$ are orthogonal, for all possible values of n, α and β subject to $\alpha \neq \beta$:

  $$\langle \Sigma_{i_n=\alpha}, \Sigma_{i_n=\beta} \rangle = 0, \qquad \alpha \neq \beta$$

  - ordering:

  $$\|\Sigma_{i_n=1}\| \geq \|\Sigma_{i_n=2}\| \geq \cdots \geq \|\Sigma_{i_n=r_n}\| > 0$$

  and

  $$\|\Sigma_{i_n=r_n+1}\| = \cdots = \|\Sigma_{i_n=I_n}\| = 0$$

  for all possible values of n.

The Frobenius-norms $\|\Sigma_{i_n=i}\|$, symbolized by $\sigma_i^{(n)}$, are the n-mode singular values of Φ and the vectors $u_i^{(n)}$ its ith n-mode singular vectors.

The SVD [6, 29] of a real $(m \times n)$-matrix A turns out to be a particular case of the Higher-Order SVD, with $n = 2$:

$$A = USV^{\mathrm{T}} = S \times_1 U \times_2 V, \qquad S \in \mathbb{R}^{m \times n}, \ U \in \mathbb{R}^{m \times m}, \ V \in \mathbb{R}^{n \times n}$$

as deduced at once from (4.77).

The HOSVD method

Let us start from factorization (4.59) of the mixing system M. Matrix B is determined from the 2nd-order statistics of the observations $\mathbf{y}$ via the PCA, as explained in section 4.5.1, whereas matrix Q will be obtained with the help of the fourth-order cumulant array of the available data. Again, the reason why third-order cumulants are not used is simply because many practical probability distributions are almost symmetric and so their cumulants of that order are usually bad conditioned. The 4th-order cumulants array of the observed signals is given by:

$$C_{ijkl}^y = \mathrm{Cum}[y_i, y_j, y_k, y_l], \qquad C_4^y \in \mathbb{R}^{p \times p p \times p \times p}. \tag{4.79}$$

From the BSS model equation (4.5) and the independence property of cumulants:

$$C_4^y = C_4^{Mx} + C_4^n, \tag{4.80}$$

since sources and noise are, by hypothesis A6, mutually independent, where C_4^y, C_4^{Mx} and C_4^n contain the fourth-order cumulants of $\mathbf{y}$, $M\mathbf{x}$ and $\mathbf{n}$, respectively. If the noise is Gaussian, from the Gaussianity property of cumulants it follows that $C_4^n = 0$, and from the multilinearity property (4.20)/(4.78):

$$C_4^{Mx} = C_4^x \times_1 M \times_2 M \times_3 M \times_4 M \qquad (4.81)$$

in which $C_4^x \in \mathbb{R}^{q \times q \times q \times q}$ and "$\times_n$" denotes the n-mode product (see previous section).

At this point, let us define the 4-dimensional array

$$\Phi = C_4^{B^*y}, \qquad (4.82)$$

which is the 4th-order cumulant array of the whitened observations $\mathbf{z}$ (or simply the sources estimated by the PCA). From (4.59), (4.81) and the corollaries 1 and 2 of the preceding section, (4.82) becomes:

$$\Phi = C_4^y \times_1 B^* \times_2 B^* \times_3 B^* \times_4 B^* =$$
$$= (C_4^x \times_1 M \times_2 M \times_3 M \times_4 M) \times_1 B^* \times_2 B^* \times_3 B^* \times_4 B^* =$$
$$= C_4^x \times_1 Q \times_2 Q \times_3 Q \times_4 Q. \quad (4.83)$$

By definition, Q is orthogonal and also C_4^x is all-orthogonal in the sense of the previous section, owing to the sources independence. Hence the last term of equation (4.83) is nothing else but the higher-order SVD of the 4th-order cumulants array Φ. Finally, in order to relieve the computational load of this method, Q may be calculated as the left singular matrix of the unfolding of Φ, defined as:

$$\Phi_{\text{unfold}}(i, j + q(k-1) + q^2(l-1)) = \Phi(i, j, k, l), \qquad i, j, k, l = 1, 2, \ldots, q. \quad (4.84)$$

Matrix Φ_{unfold} consists of q^2 matrices of dimensions $q \times q$ placed in a row in such a way that its overall dimensions are $q \times q^3$. After B and Q have been found, the linear transformation coefficients are calculated through (4.59) and, from (4.63), the source signals as $\hat{\mathbf{x}} = Q^T\mathbf{z}$. Equivalently, considering the sample matrices and the SVD-based PCA (4.56):

$$\hat{X} = \sqrt{T}\, Q^T V^T, \qquad (4.85)$$

or just Q^T times the whitened observation sample matrix Z given by the 2nd-order analysis.

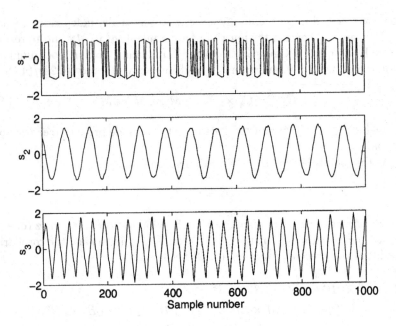

Figure 4.16: Source signals retrieved by the ICA-HOSVD method from the mixtures of figure 4.11.

Examples

<u>Synthetic signals.</u>

The sources reconstructed by the ICA-HOSVD method from the instantaneous linear mixture of figure 4.11 are displayed in figure 4.16. The separation results considerably improve, again, compared to those supplied by the PCA. The source waveforms are very similar to those obtained by the ICA-HOEVD method, although slightly worse. The difference between the results offered by these methods are further substantiated by the performance indices shown in table 4.1, page 245.

<u>Fetal ECG extraction.</u>

The performance of the HOSVD method in the FECG extraction context is condensed by the retrieved source signals represented in figure 4.17. The HOSVD presents a similar behaviour to the HOEVD, but with a slight quality decrease in the second FECG source (7th waveform in fig. 4.15 and 5th waveform in fig. 4.17). Note that the application of the HOEVD had left this signal unchanged in comparison with the PCA, but the application of the HOSVD worsens it. See table 4.2 in page 246 for the specific values of the performance indices measuring statistical independence and computational burden.

4.6.3 Extended Maximum-Likelihood (EML)

This BSS method, introduced in [66], is based on the plane-rotation notion advocated by the HOEVD procedure (section 4.6.1). Instead of maximizing a contrast function, the EML method capitalizes on the relationship between the scatter-plots of the sources and the whitened observations. After some statistical and algebraic manipulations, an estimator of the rotation angle estimator that remains to be unveiled from the whitened sensor output in the noiseless two-source two-sensor scenario is derived. Later, the reason for the acronym "EML" is justified, and links with other BSS methods established. Finally, the geometrical insights behind the method are offered. Some simulation examples illustrate the behaviour of this method.

Preliminaries

We have seen how, in the noiseless BSS problem, the source signals and the whitened observations are related through an orthogonal matrix (equation (4.63)). From now on, let us focus on the simplified 2-source 2-sensor BSS scenario, where the matrix Q becomes an elementary Givens rotation of the form:

$$Q = \begin{bmatrix} \cos\theta & -\sin\theta \\ \sin\theta & \cos\theta \end{bmatrix}. \tag{4.86}$$

The problem, then, is reduced to determining the angle θ from the whitened observations. Once this has been achieved, a counter-rotation of that angle value will provide the original source waveforms. To extend the procedure to more than two sources, the pairwise extension employed by the HOEVD can be used.

The scatter diagram.

Let us consider the pair of signals $\mathbf{z}(k) = [z_1(k), z_2(k)]^\mathrm{T}$. The bidimensional plot of the amplitude points $(z_1(k), z_2(k))$, $k = 1, 2, \ldots$, is called **scatter diagram** [30], an example of which was already displayed in figure 4.9. The scatter diagram is an approximation of the true jpdf of the corresponding pair of variables. Areas within the scatter diagram of high point density correspond to a high value for the the associated jpdf around the same area. The left panel of figure 4.18 shows the scatter diagram of

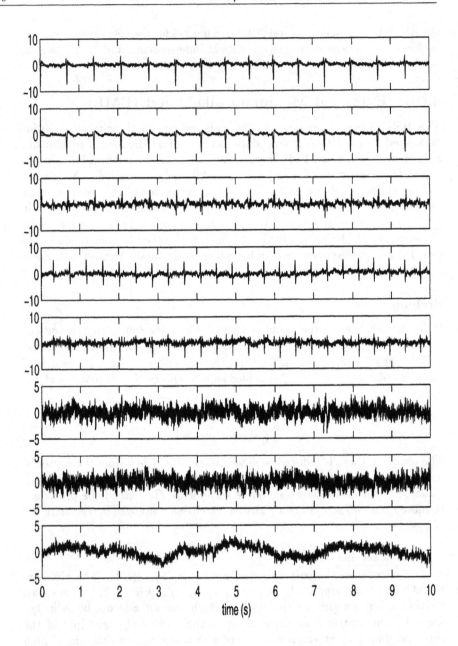

Figure 4.17: Source signals retrieved by the ICA-HOSVD method from the electrode recordings shown in figure 4.7.

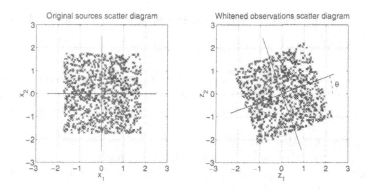

Figure 4.18: Scatter diagrams for 1000 samples of: (Left) original sources, two uniformly distributed independent signals; (Right) signals after pre-whitening.

1000 realizations of two uniformly distributed independent random variables (x_1, x_2), the source signals. It can be appreciated, in the first place, how the density is, roughly, uniform over the definition range, which means that the jpdf is flat. Moreover, it actually corresponds to a jpdf of independent variables, since it can be decomposed as the product of its marginal pdf's. The effect of an orthogonal transformation is a scatter-plot rotation. Effectively, the scatter diagram corresponding to the whitened pair of variables (z_1, z_2), connected to the sources according to expression (4.63), is represented in the right-hand panel of figure 4.18. It preserves exactly the shape of the previous one, but it is rotated an angle θ with respect to the new axes. This is directly associated with the preservation of the pdf shape, described through the pdf-transformation expression:

$$p_z(\mathbf{z}) = \frac{p_x(Q^{-1}\mathbf{z})}{|\det Q|} = p_x(Q^{\mathrm{T}}\mathbf{z}). \tag{4.87}$$

Observe that the source directions are the symmetry axes of the pre-whitened jpdf.

The scatter diagram points accept a polar as well as a complex form representation:

$$\left. \begin{array}{l} (x_1(k), x_2(k)) = x_1(k) + jx_2(k) = \rho_k e^{j\phi'_k} = \rho_k \angle \phi'_k \\ (z_1(k), z_2(k)) = z_1(k) + jz_2(k) = \rho_k e^{j\phi_k} = \rho_k \angle \phi_k \end{array} \right\} \quad k = 1, 2, \ldots \tag{4.88}$$

where, from (4.63) and (4.86), the angles ϕ'_k and ϕ_k are readily related by

$$\phi_k = \phi'_k + \theta. \tag{4.89}$$

These polar and complex form representations will prove very convenient in the the following development, and will be helpful in providing a very interesting geometrical interpretation of the estimator derived in the next section.

Some statistical relationships.

Before proceeding, let us define the following statistical terms. Note that Kendall's notation [57] for the pairwise moments and cumulants will be used hereafter. In the first place,

$$\mu^x_{mn} = \mathrm{E}[x_1^m x_2^n] \tag{4.90}$$

represents the $(m + n)$th-order moment of the bivariate random variable $\mathbf{x} = [x_1,\, x_2]^{\mathrm{T}}$. Analogously,

$$\kappa^x_{mn} = \mathrm{Cum}[x_1^m,\, x_2^n] = \mathrm{Cum}[\underbrace{x_1, \ldots, x_1}_{m}, \underbrace{x_2, \ldots, x_2}_{n}] \tag{4.91}$$

denotes the $(m + n)$th-order cumulant of the same pair of variables. Similar notation can be employed for the whitened observations, just by changing the super-index "x" by "z" in the moment and cumulant expressions. Furthermore, it will be useful to recall the following relationships [57]:

$$\begin{aligned}
\kappa^x_{40} &= \mu^x_{40} - 3\mu^{x\,2}_{20} = \mu^x_{40} - 3 \\
\kappa^x_{04} &= \mu^x_{04} - 3\mu^{x\,2}_{02} = \mu^x_{04} - 3 \\
\mu^x_{22} &= \mu^x_{20}\mu^x_{02} = 1.
\end{aligned} \tag{4.92}$$

The first two equations express the 4th-order marginal cumulants of the sources as functions of their respective moments. The assumption that both signals are unit-power has been borne in mind. These two equations are also true for the decorrelated measurements, since they are both unit-power as well. The last identity comes from the source statistical independence assumption.

A rotation angle estimator

Dropping the time index k in the sequel for convenience, let us define:

$$\xi \triangleq \mathrm{E}[\rho^4 e^{j4\phi}]\,. \tag{4.93}$$

According to relationships (4.88) and (4.89), equation (4.93) accepts an expansion as a function of the sources and the unknown rotation angle:

$$\xi = e^{j4\theta} \mathrm{E}\left[\rho^4 e^{j4\phi'}\right] = e^{j4\theta} \mathrm{E}\left[(x_1 + jx_2)^4\right]. \tag{4.94}$$

But, from the expressions given in (4.92), the above expectation turns out to be:

$$\mathrm{E}\left[(x_1 + jx_2)^4\right] = \kappa^x_{40} + \kappa^x_{04}, \tag{4.95}$$

and so ξ may be expressed as a function of the source statistics:

$$\xi = e^{j4\theta}(\kappa^x_{40} + \kappa^x_{04}). \tag{4.96}$$

From the last equation, angle θ could be estimated as:

$$\hat{\theta} = \frac{1}{4} \, \mathrm{angle}\left(\xi \cdot \mathrm{sign}(\kappa^x_{40} + \kappa^x_{04})\right), \tag{4.97}$$

where function "angle$(a + jb)$" supplies with the principal-value argument of the complex number $a + jb$, i.e., its range is the interval $]-\pi, \pi]$. The only difficulty is that the argument of the sign function is not known, because by definition the sources are not known either. However, from (4.88) and (4.92):

$$\gamma \triangleq \mathrm{E}\left[\rho^4\right] - 8 = \mathrm{E}\left[(x_1^2 + x_2^2)^2\right] - 8 = \kappa^x_{40} + \kappa^x_{04} \tag{4.98}$$

which is also available as a function of the whitened data as:

$$\gamma = \mathrm{E}\left[(z_1^2 + z_2^2)^2\right] - 8, \tag{4.99}$$

since, from (4.88), $\rho^2 = x_1^2 + x_2^2 = z_1^2 + z_2^2$. As a result, the following angle estimator arises:

$$\hat{\theta}_{\mathrm{EML}} = \frac{1}{4} \, \mathrm{angle}\left(\xi \cdot \mathrm{sign}(\gamma)\right). \tag{4.100}$$

To arrive at this expression [66] no assumptions on the source pdf's have been made at all, which makes this estimator valid for *any* source distribution combination, no matter their symmetry and tail, with the only condition that the source kurtosis sum be not zero. In such a case, the magnitude of ξ would also be null and θ could not be estimated from it any more.

Connection with other methods

In [30] an approximate ML estimator of θ is derived. To that end, the jpdf of the sources is expanded in terms of its Gram-Charlier development. This is a pdf expansion as a function of the cumulants of the random variables involved, very similar to the Edgeworth expansion [57]. Nevertheless, the nature of such an expansion makes it only valid to symmetric sources (with zero skewness) and normalized kurtosis value ranging between 0 and 4. By using this expansion on the log-likelihood of the whitened observations, and assuming that both sources have identical distributions, the following approximate ML estimator of θ is found:

$$\hat{\theta}_{\mathrm{ML}} = \frac{1}{4}\arctg\frac{\sum_k \rho_k^4 \sin 4\phi_k}{\sum_k \rho_k^4 \cos 4\phi_k}. \tag{4.101}$$

Taking into account the relationships between the "arctg($\cdot$)" and "angle($\cdot$)" functions, and the sample estimate of (4.93), it turns out that the above estimator is identical to (4.100) when the angle of its complex arguments is in the interval $]-\frac{\pi}{2}, \frac{\pi}{2}]$. When that argument lies outside that interval, a $\pm\frac{\pi}{4}$-radian bias appears in expression (4.101) relative to (4.100). Therefore, estimator (4.100) can be regarded as a generalization of estimator (4.101), which overcomes those restrictions and is valid for nearly all source distribution combination. This is the reason for the adjective "extended" in the former. The term 'ML', however, should not lead to confusion, since (4.100) may not be the ML estimator in all cases, but only (at least) under the conditions under which (4.101) was developed.

Yet there exists another interesting link between this method and some other results in the literature. It is worth computing ξ as a function of the statistical properties of the whitened observations. On the one hand, from (4.93):

$$\xi = \mathrm{E}\big[(z_1 + jz_2)^4\big] = (\kappa_{40}^z - 6\kappa_{22}^z + \kappa_{04}^z) + j4(\kappa_{31}^z - \kappa_{13}^z), \tag{4.102}$$

and, on the other hand, from (4.96), (4.98) and (4.99):

$$\xi = e^{j4\theta}\gamma = e^{j4\theta}(\kappa_{40}^z + 2\kappa_{22}^z + \kappa_{04}^z). \tag{4.103}$$

Last two equations state that θ can be determined from the 4th-order cumulants of the decorrelated measurements. In particular, the modulus of (4.102) and (4.103) must be equal, which leads to a relationship among the 4th-order cumulants of the whitened sensor outputs:

$$(\kappa_{31}^z - \kappa_{13}^z)^2 - \kappa_{22}^z(\kappa_{40}^z + \kappa_{04}^z) + 2(\kappa_{22}^z)^2 = 0, \tag{4.104}$$

This relationship was originally deduced by Comon in [22] and [21] follow-ing different and more algebraic arguments. Note, however, that Comon's estimator is essentially distinct from (4.100). For instance, the former only utilizes 4th-order cross-cumulants of the whitened signals, whereas the lat-ter employs all the 4th-order statistical information, including the marginal cumulants.

Geometrical interpretation

The apparently obscure estimator expression (4.100) accept an illuminat-ing geometrical interpretation when the two sources are symmetrically dis-tributed. Firstly, consider a mixture of two leptokurtic (also called super-Gaussian or long-tailed, i.e., with positive kurtosis value) pdf's, such as, for instance, two Laplacian distributions. Since both sources kurtosis are pos-itive, it is guaranteed that the sum is. The jpdf of the unit-power sources exhibits highest values along the lines defined by the angles $\phi' = n\pi/2$, $n = 0, 1, 2, 3$. Accordingly, the source scatter diagram has a maximum point concentration along the same lines. On the other hand, as it has already been explained in section 4.6.3, the jpdf of the whitened measurements looks like that of the sources, but rotated by θ degrees, θ being unknown. As a result, the new scatter diagram displays the highest density along $\phi = \theta + n\pi/2$, $n = 0, 1, 2, 3$. Now, assume that all the scatter diagram points are trans-formed according to

$$\rho_k e^{j\phi_k} \mapsto \rho_k^4 e^{j4\phi_k}. \tag{4.105}$$

In particular, all the previous points on $\phi = \theta + n\pi/2$ will now lie along $\phi = 4\theta$, as graphically depicted in figure 4.19. In this way it can be argued that a *centroid* calculated as the resultant mean point after the transformation, that is

$$\xi = \mathrm{E}\left[\rho_k^4 e^{j4\phi_k}\right], \tag{4.106}$$

will show that orientation. As a result, (4.100) will provide the required angle with which the whitened jpdf is rotated with respect to the true source jpdf. It is interesting to observe that the exponent of ρ_k in the transformation (4.105) could be, in principle, arbitrary: ideally the centroid orientation is independent of such an exponent, at least for a number of samples high enough.

In the second place, let us now consider two platykurtic (also named sub-Gaussian or short-tailed, i.e., with negative kurtosis value) distributions playing the role of sources. Hence, their kurtosis sum is negative. For in-stance, two sinusoidal signals. In this case, the maximum concentration in the

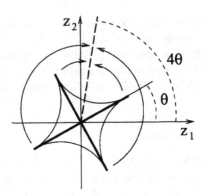

Figure 4.19: Centroid location and angle estimation for a mixture of two leptokurtic distributions.

source scatter diagram occurs along the lines $\phi' = \pi/4 + n\pi/2$, $n = 0, 1, 2, 3$. Therefore, in the scatter diagram of the signals after decorrelation this high density will be shown along $\phi = \theta + \pi/4 + n\pi/2$, $n = 0, 1, 2, 3$. These points will all gather at $\phi = 4\theta + \pi$ when transformed according to (4.105) and so the centroid (4.106) will also have this orientation (figure 4.20). In conclusion, the centroid projected across the origin, i.e., $-\xi$, forms an angle 4θ with respect to the x-axis. That is the reason for the "sign($\cdot$)" term in the estimator expression (4.100). It is precisely the introduction of this simple extra term that allows to extend the validity domain of the approximate ML estimator (4.101).

For combinations of platykurtic and leptokurtic distributions, the geometrical interpretation becomes more involved and less intuitive, but the results obtained in the previous section indicate that those cases can also be explained in terms of the above two instances, comprising sources with positive and negative kurtosis sum.

Examples

Synthetic signals.

Figure 4.21 shows the source signals obtained by the ICA-EML method, from the measurements represented in figure 4.11, while the resultant performance indices appear in the last row of table 4.1. For this particular mixture realization, the ICA-EML obtains the best results out of the four compared procedures. A more exhaustive comparison among all these four BSS methods is carried out in section 4.7.

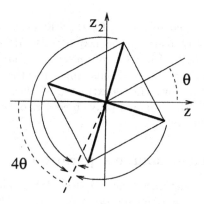

Figure 4.20: Centroid location and angle estimation for a mixture of two platykurtic distributions.

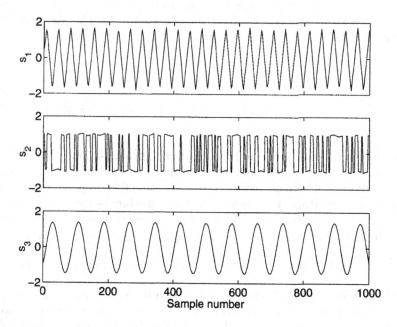

Figure 4.21: Source signals recovered by the ICA-EML method from the mixtures of figure 4.11.

Fetal ECG extraction.

Regarding our particular dataset for the FECG extraction problem, the EML obtains an almost identical separation to the HOEVD, as can be appreciated in figure 4.22. Even the reconstructed sources are arranged in the same order. As a measure of this similarity, the mean square difference between the source waveforms extracted by these two methods is just $\varepsilon^2 = 0.0054 = -22.7\,\mathrm{dB}$. Two qualitative measures of the separation performance appear in table 4.2.

4.6.4   Equivariant source separation

The multiplicative structure of the noiseless BSS model (4.54) brings out into play the notion of **equivariance**. An estimator $\widehat{\mathsf{M}}$ of M is said to be equivariant when [12, 13]:

$$\widehat{\mathsf{M}}(A\mathbf{y}) = A\widehat{\mathsf{M}}(\mathbf{y})J, \qquad \forall A \text{ invertible}, \quad J \text{ quasiidentity}, \qquad (4.107)$$

that is, when transformations in the observed data produce analogous transformations in the estimated parameter. Now, consider the global matrix obtained through an equivariant estimator of the mixing matrix:

$$G = WM \underset{\substack{\uparrow \\ W=\hat{M}^\bullet}}{=} \widehat{\mathsf{M}}(\mathbf{y})^* M \underset{\substack{\uparrow \\ \text{BSS model (4.54)}}}{=} \widehat{\mathsf{M}}(M\mathbf{x})^* M =$$

$$\underset{\substack{\uparrow \\ \text{equivariance}}}{=} (M\widehat{\mathsf{M}}(\mathbf{x}))^* M = \widehat{\mathsf{M}}(\mathbf{x}). \quad (4.108)$$

That is, the global matrix, and hence the source signals, obtained through an equivariant estimator of the mixing matrix do not depend at all on the mixing matrix itself, but only on the sources. This attractive property enjoyed by equivariant algorithms is known as **uniform performance**. An estimator fulfilling (4.107) is also referred to it as *fully invariant*. Along the same lines, an estimator $\widehat{\mathbb{Q}}$ of the orthogonal matrix Q is called *orthogonal invariant* if

$$\widehat{\mathbb{Q}}(A\mathbf{z}) = A\widehat{\mathbb{Q}}(\mathbf{z})J, \qquad \forall A \text{ invertible}, \quad J \text{ quasiidentity}. \qquad (4.109)$$

In the noiseless case, orthogonal invariant algorithms are fully invariant [12].

The question now is if such algorithms exist and, in the case they do, how to obtain them. This is answered in [13]. It is shown that there is actually a wide variety of equivariant BSS methods. Among them, methods based on maximum-likelihood estimation as well as procedures relying on contrast functions are easily proved to be equivariant. In the next section we will

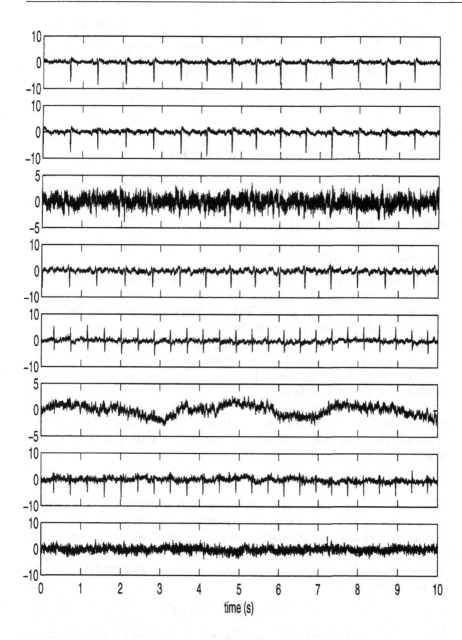

Figure 4.22: Source signals retrieved by the ICA-EML method from the electrode recordings shown in figure 4.7.

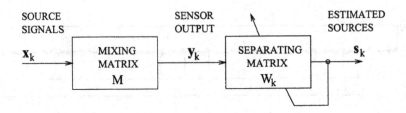

Figure 4.23: Adaptive BSS.

see how an adaptive separation algorithm may acquire this property, which requires the introduction of two important notions: serial updating and the relative gradient [13,14,17,37].

Equivariant adaptive algorithms

Off-line or batch-processing methods obtain an estimate of the separating matrix by processing a whole block or batch of sensor-output samples, as explained in section 4.5.1. All the BSS procedures presented thus far belong to this kind of techniques. Nevertheless, certain application may require to perform the source extraction and mixing matrix identification in real time, i.e., in a sample-by-sample fashion. In such a case, the separation has to be carried out adaptively. The typical set-up for adaptive, or on-line, blind separation is sketched in figure 4.23.

The uniform performance characteristic described in the previous section can be retained in adaptive algorithms through the device of serial updating. A **serial updating** rule for the adaptive estimation of the separating matrix reads:

$$W_{k+1} = W_k - \lambda_k H(\mathbf{s}_k) W_k, \tag{4.110}$$

where $H(\cdot)$ is a vector-to-matrix mapping such that $\mathrm{E}[H(\mathbf{s})] = 0$ when a the output vector $\mathbf{s}$ contains a valid separation solution. It is called 'serial' because the separating matrix at iteration $(k+1)$ is obtained by plugging the matrix $(I - \lambda_k H(\mathbf{s}_k))$ into the separating matrix estimated at the previous iteration. The above equation corresponds to an analogous updating for the global system:

$$G_{k+1} = G_k - \lambda_k H(G_k \mathbf{x}_k) G_k. \tag{4.111}$$

Note that the trajectories of the global matrix depend on the sources and on its initial value G_0. However, the evolution of the global system is independent of the mixing matrix M, in the sense that changing the mixing matrix is tantamount to changing the initial value of the separator, W_0. In conclusion, a separation algorithm with serial updating exhibits uniform performance.

Let us assume that the separation is accomplished by the minimization of a cost function $\phi(W)$. Before proceeding, some definitions are in order. First, we denote $\langle \cdot \mid \cdot \rangle$ the Euclidean scalar product of matrices:

$$\langle A \mid B \rangle = \text{trace}[A^T B], \qquad \langle A \mid A \rangle = \|A\|_{\text{Fro}}^2. \tag{4.112}$$

The gradient of ϕ at W is the $q \times p$ matrix denoted by $\phi'(W)$ such that the first-order expansion of ϕ at W reads:

$$\phi(W + \varepsilon) = \phi(W) + \langle \phi'(W) \mid \varepsilon \rangle + o(\varepsilon). \tag{4.113}$$

The **relative gradient** of ϕ at W is the $q \times q$ matrix denoted by $\nabla \phi(W)$ such that:

$$\phi(W + \varepsilon W) = \phi(W) + \langle \nabla \phi(W) \mid \varepsilon \rangle + o(\varepsilon). \tag{4.114}$$

Combining the last two expansions with (4.112) yields: $\nabla \phi(W) = \phi'(W)W^T$. In a similar way to other gradient-based rules, a relative gradient rule consists in aligning the relative variation ε in a direction opposite to the relative gradient, that is, in modifying W into $(W + \varepsilon W)$, with $\varepsilon = -\lambda \nabla \phi(W)$. In such a case, the cost function becomes:

$$\phi(W - \lambda \nabla \phi(W)) = \phi(W) + \langle \nabla \phi(W) \mid -\lambda \nabla \phi(W) \rangle + o(\lambda \nabla \phi(W)) =$$
$$= \phi(W) - \lambda \|\nabla \phi(W)\|_{\text{Fro}}^2 + o(\lambda), \quad (4.115)$$

so for small enough positive values of the adaption coefficient λ, the objective function ϕ is decreased, as long as $\nabla \phi(W) \neq 0$. Consequently, the relative gradient rule is consistent with the serial updating (4.110):

$$W_{k+1} = W_k - \lambda_k \nabla \phi(W_k) W_k. \tag{4.116}$$

Adaptive algorithm (4.110) is entirely characterized by the function $H(\cdot)$. In order to arrive at a more precise structure for (4.110), let us focus on a cost function of the form:

$$\phi(W) \triangleq \text{E}[f(\mathbf{s})] = \text{E}[f(W\mathbf{y})]. \tag{4.117}$$

Its relative gradient can be determined as:

$$\nabla \phi(W) = \text{E}\left[\mathbf{f}'(\mathbf{s})\mathbf{s}^T\right], \tag{4.118}$$

where $\mathbf{f}'(\mathbf{s})$ is the gradient of f at $\mathbf{s}$, $\mathbf{f}'(\mathbf{s})_i = \partial f(\mathbf{s})/\partial s_i$. By deleting the expectation operation, a stochastic relative gradient rule is obtained:

$$W_{k+1} = W_k - \lambda_k \mathbf{f}'(\mathbf{s}_k) \mathbf{s}_k^T W_k, \qquad (4.119)$$

where we can identify $H(\mathbf{s}) = \mathbf{f}'(\mathbf{s})\mathbf{s}^T$. Observe that the stochastic optimization of the same cost function by means of the standard gradient leads to the adaption rule:

$$W_{k+1} = W_k - \lambda_k \mathbf{f}'(\mathbf{s}_k) \mathbf{y}_k^T, \qquad (4.120)$$

which does not enjoy uniform performance properties.

When the contrast ϕ has to be minimized under the decorrelation constraint $R_s \triangleq \mathrm{E}[\mathbf{ss}^T] = I_q$, it is named **orthogonal contrast**. An example of a batch procedure for optimizing an orthogonal contrast function is the HO-EVD method described in section 4.6.1. If ϕ in (4.117) is to be minimized under such a decorrelation convention, then the following function $H(\cdot)$ is obtained [13, 17, 37]:

$$H(\mathbf{s}) = \mathbf{ss}^T - I + \mathbf{g}(\mathbf{s})\mathbf{s}^T - \mathbf{sg}(\mathbf{s})^T, \qquad (4.121)$$

with $\mathbf{g}(\mathbf{s}) = \mathbf{f}'(\mathbf{s})$. This family of algorithms is known as *EASI*, for *Equivariant Adaptive Separation via independence*
. A particular algorithm is constructed by defining a vector-to-vector function of the outputs $\mathbf{g}(\mathbf{s})$. This function, however, must be selected with care. It must be such that the expected value of $H(\cdot)$ in (4.121) vanishes when W is a valid separating matrix, that is, it must make a valid separating matrix a stationary point of the algorithm. For instance, a component-wise function $\mathbf{g}(\mathbf{s}) = [g_1(s_1), \ldots, g_q(s_q)]^T$ fulfils this condition for any q arbitrary non-linear functions $g_i(s_i)$. On the other hand, the stability and asymptotic performance of the algorithm also depends on the choice of these non-linear functions with respect to the source distribution. As an example, for cubic non-linearities $g_i(s_i) = s_i|s_i|^2$ the stability condition dictates that the sum of normalized kurtosis of every pair of sources must be negative [17].

Examples

For sinusoid, triangular and binary sources, the normalized kurtosis values are, respectively, -1.5, -1.2 and -2, and therefore the choice of cubic non-linearities fulfils the aforementioned convergence condition. By using the mixture samples of figure 4.11, along with cubic non-linearities and a fixed adaption step size $\lambda = 5 \times 10^{-3}$, the evolution of the global matrix coefficients are obtained by the EASI method as shown in figure 4.24. Three of such

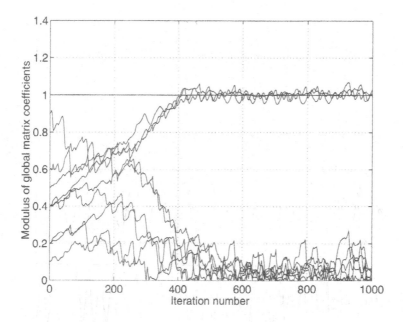

Figure 4.24: Trajectories of the global system elements for the separation of the mixture in figure 4.11 by the ICA-EASI method, with cubic non-linearities and $\lambda = 5 \times 10^{-3}$.

elements converge (in absolute value) to 1, whereas the rest do so to 0. Hence, a good separation solution is achieved. The corresponding source signals recovered by this method appear in figure 4.25. Observe how after the transient period (which lasts, approximately, until the 400th sample), the adaptively estimated source signals resemble the original sources.

4.7 Comparison

This section is entirely devoted to assessing in finer detail the performance of the BSS methods theoretically outlined in the preceding sections. With the help of a number of Monte Carlo simulations on computer generated data the methods' performance is evaluated in various contexts, such as varying sample size and varying signal-to-noise ratio in a Gaussian additive noise environment. All experiments are carried out on the MATLAB environment.

The followed methodology is reported in section 4.7.1. The results are

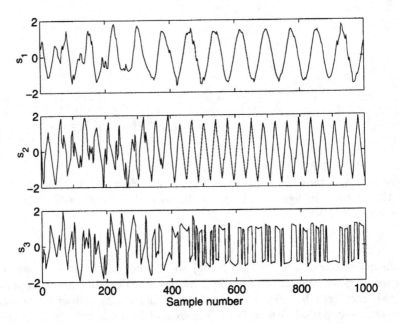

Figure 4.25: Source signals recovered by the ICA-EASI method from the mixtures of figure 4.11, with cubic non-linearities and $\lambda = 5 \times 10^{-3}$.

discussed in section 4.7.2.

4.7.1 Experimental methodology

Initially, a BSS scenario is selected, composed of:

- Number and type of source signals.

- Number of samples per signal, or sample size, T.

- Noise conditions, including noise distribution and signal-to-noise ratio (SNR).

The SNR is defined on a sensor-wise basis:

$$\text{SNR}_i = \left. \frac{\text{Power due to sources}}{\text{Power due to noise}} \right|_{i\text{th sensor}} \quad (4.122)$$

and is chosen to be the same for all sensor outputs, $\text{SNR}_i = \text{SNR}$, $1 \leqslant i \leqslant p$. This means that the noise power depends on the particular sensor, according to the elements of the mixing matrix. In particular, the power of the noise signal at the ith sensor output is thus given by

$$\text{SNR} = \frac{\sum_{j=1}^{q} m_{ij}^2 P_{x_j}}{P_{n_i}} \quad \Rightarrow \quad P_{n_i} = \frac{\|\mathbf{m}_{i:}\|^2}{\text{SNR}}, \quad (4.123)$$

$\mathbf{m}_{i:}$ denoting the ith row of the mixing matrix M, P_{n_i} the ith-noise power and P_{x_j} the jth-source power, assumed to be unity. Hence, given a mixing matrix and particular unit-power realizations of source and noise signals, the observations are built up as

$$\mathbf{y}(k) = M\mathbf{x}(k) + \Delta_n \mathbf{n}(k), \quad (4.124)$$

where

$$\Delta_n = \sqrt{\text{diag}(P_{n_1}, \ldots, P_{n_p})} = \sqrt{\text{diag}\left(\text{diag}\left(MM^{\mathrm{T}}\right)\right)/\text{SNR}} =$$
$$= 10^{-(\text{SNR (dB)}/20)} \sqrt{\text{diag}\left(\text{diag}\left(MM^{\mathrm{T}}\right)\right)}. \quad (4.125)$$

The coefficients of the mixing matrix M are drawn from a random variable uniformly distributed between -1 and 1. The number of sensors is taken to coincide with the number of sources, $p \equiv q$. After mixing, the sensor output realization is fed into the methods (the same realization is used for all of them), and after the separations are obtained, the values for the performance indices depicted in section 4.3 are worked out. This is repeated, with different mixture realizations but preserving the initially chosen conditions, over ν Monte Carlo runs, where $\nu T = 5 \cdot 10^5$. Finally, at the end of all repetitions the mean value of the parameters are computed, and displayed in figures 4.26 to 4.35.

Two main scenarios are tested:

- Noiseless scenario (SNR $= \infty$), with varying sample size ranging from 100 to 10000 samples (figures 4.26 – 4.30). Specifically, the sample sizes considered are 100, 500 and from 1000 to 10000 in 1000-sample steps.

- Gaussian noise scenario, with fixed sample size (T $= 5000$ samples) and varying SNR from -30 to 30 dB, in 5 dB steps (figures 4.31 – 4.33).

The purpose of the first simulation is to assess the methods' performance as a function of the number of processed samples, whereas the other one evaluates how the methods respond at different SNRs in Gaussian-noise corrupted measurements. Each of those conditions could be evaluated by means of different source distribution sets, but for simplicity we will adhere to the combination "sinusoid–triangular–binary", that has been dealt with in the previous simulation examples.

For the noisy simulations, only parameters ε^2, ISR and Υ_4 are computed. In this context indicator ϵ seems to show an inconclusive behaviour. On the other hand, computation of the number of flops is pointless, since by construction of the algorithms this parameter is not affected by the SNR.

Finally, in a noiseless environment of 5000-sample uniformly distributed sources, the computational complexity as the number of signals increases is obtained (figures 4.34– 4.35). All the results are discussed in the next section.

4.7.2   Results and discussion

Figures 4.26 to 4.30 show the results obtained for the noiseless simulation, i.e., the performance indices as a function of the sample size T. From the figures corresponding to parameters ε^2, ϵ, ISR and Υ_4, it is seen how all the HOS-based methods improve the source extraction as T increases. The reason for this general improvement is mainly that as the sample size increases the cumulant estimation becomes more accurate, and so does the separation achieved. By contrast, the PCA method does not provide a satisfactory separation, which is manifested in poor parameter values and these do not seem to depend on the number of samples processed. This behaviour was expected from the theoretical considerations of section 4.5.2. The HOEVD and the EML clearly show the best performance out of the four tested procedures. Indeed, both methods show identical results for these simulation conditions, although the latter is computationally heavier. This is due to the complex-number calculations required to obtain the centroid locations in that method. However, the EML computational burden becomes the same as HOEVD when the angle estimates are calculated as a function of the whitened sensor output cumulants. The PCA is the simplest, but at the expense of very poor performance. Next comes the HOSVD in separation quality, even though its computational cost is higher than HOEVD. The complexity increases with a linear trend for all four methods, which means that flops are a linear function of T.

Figures 4.31 to 4.33 plot the results obtained for the noisy simulations, for different SNRs. First of all, a consistent behaviour is observed for indicator ε^2. As the SNR increases, the methods get better estimations of the source signals, and so ε^2 tends to zero (except, again, for the PCA method, which gets an asymptotic value of -4 dB, that cannot be regarded as sufficient for a good waveform-preserving separation). However, when the SNR decreases, the noise signals become dominant power contributors at the sensor outputs, and as a result the algorithms 'see' them as the actual sources to be extracted. Consequently, at low enough SNRs, the sources estimated are actually the noise signals, which, moreover, are independent of the sources. This is the reason why ε^2 tends to 3 dB (2 in linear scale) as the SNR tends to minus infinity. This trend can be appreciated in figure 4.31. In that plot, as the SNR increases from 0 dB a fall of slope $-1/2$ is also observed for the HOS-based methods. Results for the HOEVD and the EML are, once more, virtually identical, followed by the HOSVD.

Results for parameter ISR reveal more information than for the noiseless case. In figure 4.32, the EML method demonstrates a slightly better performance for this parameter than the HOEVD, with the HOSVD and the PCA behind. Interestingly, the asymptotic slope of the curves of the EML and the HOEVD are the same as that obtained for index ε^2 (figure 4.31) by these two methods, i.e., around $-1/2$. The slope of the HOSVD is less steep.

Figure of index Υ_4 is more discriminating among the HOS-based methods. Its shape also admits a statistical explanation. Again, when the SNR is high a good independent-source estimation is obtained, and therefore the parameter tends to one. When the SNR is low, however, the methods begin to extract the noise, rather than the true sources, and any quality indicator computed from the separator outputs actually corresponds to the noise signals. As the noise is Gaussian one would expect to have an undetermined value of Υ_4 since ideally all noise 4th-order cumulants would vanish. The finite value obtained in figure 4.33 is owing to the cumulant estimation error caused by the finite sample size. If the noise were not Gaussian, the recovered noise signals would yield an asymptotic value of one for Υ_4 ($-\infty$ for $(1 - \Upsilon_4)$ in dB), since in such conditions the noise 4th-order cumulants would not be zero. From this figure, the HOEVD exhibits the best results, followed very closely by the EML, and the HOSVD. Observe too how the mixing decreases the degree of statistical independence of the signals, and how the application of the BSS methods gets it back. This feature is also illustrated by figure 4.29 in the noiseless case.

Finally, figures 4.34 and 4.35 show how the methods' computational complexity varies as the number of signals to be processed increases, for a fixed sample size of 5000. It can be appreciated how the HOSVD is the most expensive out of the four methods, followed by the EML, the HOEVD, and the PCA. This trend becomes apparent from 5 processed signals. As commented above, EML complexity can be made the same as HOEVD if the centroids are worked out from the cumulants.

In conclusion, from this simulation results the HOEVD and the EML exhibit the best overall performance, with an outstanding separation quality. As a matter of fact, both seem to offer the same asymptotic performance under the tested simulation conditions. The HOSVD comes always behind the other two HOS-based procedures in separation quality. The PCA, as a second-order technique, is not able to perform an adequate source waveform extraction.

4.8   Comments on the literature

BSS has been thoroughly studied since 1986. Jutten's pioneering work [33,34] provides one of the first significant approaches to this problem. The proposed solution was based on a neural network implementation (denoted by *HJ network*), but no theoretical explanation was available at that time. The neuro-algorithm aims at the cancellation of certain non-linear odd functions of the separator outputs, which implicitly introduce higher-order moments of the output components. In [23] and [26] the algorithm is explained in statistical terms, and is found to be able to separate the sources if they have non-Gaussian and symmetric pdf's. Moreover, in some pathological cases the algorithm may fail to give a proper solution. The stability analysis of the HJ network is accomplished in [56].

Lacoume and Ruiz propose an exhaustive search of the absolute minima of a response function built up from the cumulants of the Fourier transform of the data signals [35]. Constrained optimization (e.g., via the Fletcher-

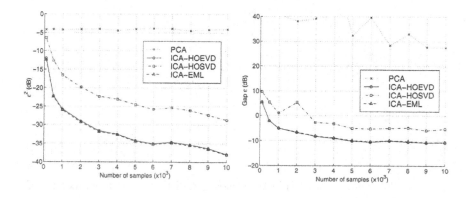

Figure 4.26: Parameter ε^2 for the separation of three noiseless mixtures of sinusoid–binary–triangular sources.

Figure 4.27: Parameter ϵ for the separation of three noiseless mixtures of sinusoid–binary–triangular sources.

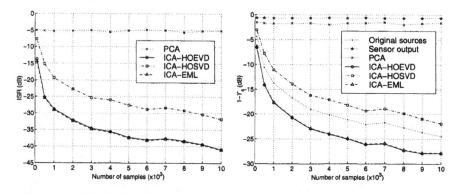

Figure 4.28: Parameter ISR for the separation of three noiseless mixtures of sinusoid–binary–triangular sources.

Figure 4.29: Parameter Υ_4 for the separation of three noiseless mixtures of sinusoid–binary–triangular sources.

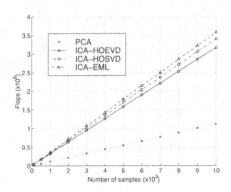

Figure 4.30: Number of flops in the separation of three noiseless mixtures of sinusoid–binary–triangular sources.

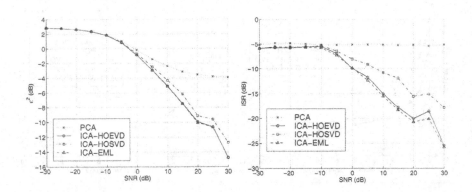

Figure 4.31: Parameter ε^2 for the separation of sinusoid–binary–triangular sources in Gaussian noise, $T = 5 \times 10^3$ sample/signal.

Figure 4.32: Parameter ISR for the separation of sinusoid–binary–triangular sources in Gaussian noise, $T = 5 \times 10^3$ sample/signal.

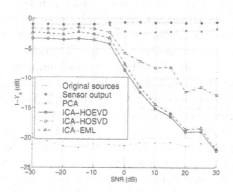

Figure 4.33: Parameter Υ_4 for the separation of sinusoid–binary–triangular sources in Gaussian noise, $T = 5 \times 10^3$ sample/signal.

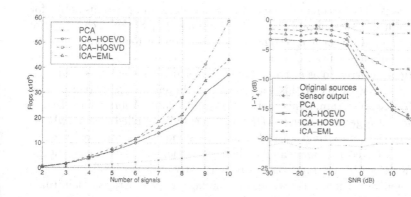

Figure 4.34: Number of flops versus number of signals, $T = 5 \times 10^3$ sample/signal.

Figure 4.35: Logarithm of the number of flops versus number of signals, $T = 5 \times 10^3$ sample/signal.

Powell algorithm) is resorted in [36]. This time the cost function to minimize consists in the sum of squares of the 4th-order pairwise cross-cumulants of the outputs. The output signals must also be uncorrelated, which becomes the constraint, although no source variance normalization is imposed. For the noiseless two-source two-sensor scenario, Mansour and Jutten propose as a cost function a particular 4th-order cross-cumulant [40]. In order to avoid spurious solutions, uncorrelation has also to be forced at the outputs, so a system of two non-linear equations turns up. The computational burden of these optimization-based procedures is more than apparent. In a bid to relieve this complexity, it is shown in [45] that the last system can be considerably simplified and the numerical search spared by applying a previous decorrelation processing to the sensor-output signals, which endorses the convenience of a two-stage approach to BSS [65]. This result relies mainly on [22] and [21], where a closed-form formula for the two-signal case is developed, giving a direct solution when the observed signals come from an orthogonal transfer matrix operating on a set of equal-power sources.

HOS is combined with matrix algebra by Cardoso in [11], where the separation is attained through the EVD of a matrix made up of particular fourth-order moments. This technique, the *Fourth-Order Blind Identification (FOBI)* algorithm, is able to recover the sources when they have different non-Gaussian symmetric pdf's, but fails in other cases. In [10] the same author improves the method by resorting to the fourth-order moments tensor of the observations. This second version of Cardoso's algorithm performs better for any set of non-Gaussian probability distributions. In [18] the eigen-structure of the cumulant tensor is investigated and the relevant fourth-order statistical information (a subset of cross-cumulants) is contained in a reduced set of matrices, referred to as eigen-matrices. The sources are extracted through a joint diagonalization of this set, which can be efficiently carried out by a Jacobi-like algorithm. This method, so-called *Joint Approximate Diagonalization of Eigen-matrices (JADE)*, is successfully applied to narrowband beamforming. Interestingly, JADE offers the same asymptotic performance as Comon's HOEVD (commented in detail in section 4.6.1), while having the advantage that its cost function can be efficiently optimized [13].

The generalization of these ideas to the higher-order array domain (naturally brought in by the multilinearity property of cumulants which allows them to be considered as tensors; refer to section 4.6.2) is carried out by De Lathauwer and colleagues. In [39] they extend the SVD to higher-order arrays — yielding the higher-order SVD — and apply it to the BSS problem, as expounded in section 4.6.2. The cumulant array of the whitened observations can be diagonalized with this new algebraic tool, thus providing a set of independent signals. The method tackles the fetal ECG problem with success [38].

For the noiseless two-source two-sensor scenario a direct solution is achieved by rooting a fourth-order polynomial equation using only fourth-order cumulants if the sources are non-Gaussian [41]. Under identical conditions, that solution is further simplified in [20], where the separating matrix is estimated by rooting only a second-degree polynomial equation involving fourth-order cumulants. Both procedures are quite simple and their accuracy is only limited by the precision of the cumulant estimation. Their major drawback is that, in order to fix the source-amplitude ambiguity, this kind of procedures often normalize the separating matrix (for instance, taking its diagonal elements equal to 1). This leads to algorithms whose behaviour depend strongly on the mixing matrix, hence renouncing the uniform performance property [13].

The concept of ICA is introduced in the context of the BSS problem by Comon in [24], as explained in section 4.6.1. The notion of ICA is intimately related to the concept of contrast function, which, as has also been seen in section 4.6.1, are functions of the output pdf whose maximization supply with independent components at the output of the separator. Another form of implementing the ICA may be found in [49]. Its contrast function is based on the Kullback-Leibler divergence and the kernel estimate of the source probability densities. Other cost functions derived earlier in the literature are also found to be contrasts, such as JADE's cost function [13, 16].

Waveform-preserving identifiability issues are discussed at length in [59] and [60]. Some identifiable BSS models (both with uncorrelated sources and with independent sources) are presented and two practical blind identification algorithms developed. The first of them is called *EFOBI (Extended FOBI)* and is a generalization of Cardoso's FOBI method which does not ignore the noise effects, but still it only works for sources with different kurtosis values. The other method — *Algorithm for Multiple Unknown Signals Extraction (AMUSE)* — exploits the second-order statistics of the source processes, when they are not temporally white. In order for this method to work properly, the sources must possess different autocorrelation at some time lag, i.e., different spectral content. In [58], Tong *et al.* prove that the minimum-variance unbiased estimates of the source signals can be obtained if and only if the transmission channel (the transfer matrix) can be blindly identified. In turn, the channel can be blindly identified if and only if there is no more than one Gaussian source. Nevertheless, if the temporal information of the observed processes is employed, the channel may be identifiable even if the sources are all Gaussian when the second-order statistics of the observation signals (such as correlation at different times) are used. In addition, the problem of channel identification is shown to be equivalent, from an algebraic point of view, to simultaneously diagonalizing the observation cumulant matrices (unfolded representations of the corresponding cumulant arrays). In [9] a general approach to the problem is given, which provides a uniform framework for the fundamental theory and covers the existing results as special cases. Two important concepts in the BSS problem are studied: separability and separation principles. *Separability* refers to the conditions that the measured signals have to satisfy for the sources to be identifiable. It turns out that separability depends only on the structure of the mixing matrix. *Separation principles* correspond to the conditions that can be applied to verify whether a successful separation has been carried out. Basically, if the signals at the output of the separator are pairwise independent, they must be the sources, as long as there are at most one Gaussian source signal. This outcome entirely agrees with identifiability conditions developed earlier in the literature.

A ML approach is undertaken in [28]. With the help of the Gram-Charlier expansion of the sources jpdf, a sub-optimal two-step procedure is developed: first, decorrelation; second: 4th-order independence. However, the cost function to be maximized in the second step is identical to the contrast function derived separately by Comon in [24]. In [50], the ML principle leads to estimating equations which depend on non-linear functions of the observations. The optimal non-linearities are the derivatives of the logarithm of the source pdf's. The ML standpoint is also adopted in [30], where by using certain simplifications a closed-form solution for the angle to be estimated in the noiseless two-source two-sensor BSS set-up is obtained. This estimator was generalized in [66] as explained in section 4.6.3. In [5] the ML principle is combined with the EM (Expectation-Maximization) algorithm to yield a method able to separate more sources than sensors when the former obey a discrete distribution.

Along the lines of the JADE and AMUSE algorithms, and based on the identifiability criteria developed in [60] and [58], a *Second-Order Blind Identification (SOBI)* algorithm is reported in [4]. It exploits the time coherence of the source signals and it consists in a joint diagonalization of several whitened covariance matrices at different time lags. Robustness is significantly increased at low additional cost by processing such a matrix set rather than a unique matrix as in AMUSE. Furthermore, the SOBI method allows — in contrast to higher-order cumulant techniques — the separation of Gaussian sources. Nevertheless, the spectral separation of the sources is still essential for a good performance of the method.

The topic of adaptive source separation is addressed in detail by Cardoso and Laheld in [17]. They derive the EASI algorithms, as depicted in section 4.6.4. Other adaptive BSS methods appear in [22] and [31]. The former is the adaptive version of the block method described therein and in [21], and basically reduces to the adaptive cumulant estimation. The latter procedure is an adaptive implementation of Gaeta-Lacoume's [28] and Comon's [24] block methods.

Many procedures for BSS resort to the 4th-order statistical information of the observed data to achieve the separation, as in [17, 22, 24, 30, 40, 66], etc. As a consequence, their performance is somehow dependent on the source statistics at that order. In [67] an interesting result is derived and its effects investigated: a simple binary distribution exhibits all possible range of normalized kurtosis values when the probability of its two events is varied accordingly. This outcome suggests that any sources may be substituted, as far as their kurtosis value is concerned, for binary distributions of appropriate parameters in order to test the asymptotic performance of such BSS methods. This hypothesis is successfully checked out with the adaptive method of [17].

For the more general case in which the mixing system is formed by unknown linear time-invariant (LTI) filters, signal separation criteria are developed in [62], [63] and [64]. In [62], only the second-order independence (uncorrelation) of the sources is exploited through a decorrelation-based technique, whereas [63] relies on their cross-polyspectra, thus resorting to the HOS of the data. It is proved again that statistical independence between the source signals is a sufficient condition for signal separation, but even weaker sufficient conditions involving their cross-polyspectra are found. A unified framework for many of these methods is presented in [64], which also analyzes their statistical stability for finite sample size.

An apparently new information theory based approach attracted a lot of interest, the so-called *information-maximization (infomax)* principle [3]. In short, the application of this principle to source separation consists in maximizing a particular contrast function: an output entropy. However, it is proved in [15] that, regarding source separation, infomax is equivalent to ML.

Symbol	Meaning
x_i	ith source signal
y_i	ith sensor output (observed signal)
n_i	ith noise signal
z_i	ith whitened sensor output
s_i	ith separator output
$\hat{x}_i$	ith estimated source signal
$\tilde{x}_i$	ith normalized source signal
$\mathbf{x}$	source signal vector
$\mathbf{y}$	sensor signal vector
$\mathbf{n}$	noise vector
$\mathbf{z}$	whitened observation vector
$\mathbf{s}$	separator output vector
$\hat{\mathbf{x}}$	estimated source vector
$\tilde{\mathbf{x}}$	normalized source vector
X	source signal matrix
Y	sensor output matrix
N	noise matrix
Z	whitened observation matrix
S	separator output matrix
$\hat{X}$	estimated source matrix
$\tilde{X}$	normalized source matrix
q	number of sources
p	number of observations
T	number of samples
M	mixing matrix
m_{ij}	mixing matrix element
$\mathbf{m}_j$	transfer vector/source direction
W	separating matrix
w_{ij}	separating matrix element
$\mathbf{w}_j$	separating vector
G	global matrix (WM)
g_{ij}	global matrix element
P	permutation matrix
D	invertible diagonal matrix
I_n	$n \times n$ identity matrix
$(\cdot)^{\mathrm{T}}$	transposition operator
$(\cdot)^*$	pseudoinverse operator

Notation and terminology used in this chapter.

Symbol	Meaning
R_x	spatial covariance matrix
$H_{ij}\{\cdot\}$	transformation
h_{ij}	linear time-invariant system
δ	Dirac's delta function
$E[\cdot]$	expected value
μ	moment
$\text{Cum}[\cdot], \kappa$	cumulant
$\text{cov}[\cdot]$	covariance
$C^x_{i_1\ldots i_n}$	nth-order cumulant array
ε^2	waveform similarity index
ϵ	gap between true and estimated mixing matrix
Υ	statistical independence index
ϱ_{ij}	rejection rate between ith and jth source
ISR	interference-to-signal ratio
SNR	signal-to-noise ratio

Notation and terminology used in this chapter (cont.).

PARAMETER → METHOD ↓	ε^2	ϵ	ISR	Υ_4	Flops
PCA	9.96×10^{-2} (-10.0)	2.32 (3.7)	56.1×10^{-3} (-12.5)	0.64 (-4.5)	1.41×10^5
ICA-HOEVD	0.14×10^{-2} (-28.5)	0.13 (-9.0)	0.7×10^{-3} (-31.6)	0.99 (-20.5)	3.56×10^5
ICA-HOSVD	0.90×10^{-2} (-20.5)	0.50 (-3.0)	4.5×10^{-3} (-23.5)	0.97 (-14.9)	4.04×10^5
ICA-EML	0.12×10^{-2} (-29.2)	0.10 (-9.8)	0.6×10^{-3} (-32.1)	0.99 (-20.4)	3.87×10^5

Table 4.1: Performance indices for the separation of the mixture of figure 4.11. In parentheses, values in dB.

METHOD → PARAMETER ↓	PCA	ICA-HOEVD	ICA-HOSVD	ICA-EML
Υ_4	0.32 (−1.69)	0.85 (−8.25)	0.64 (−4.45)	0.84 (−7.86)
Flops	4.1×10^6	18.5×10^6	28.2×10^6	21.3×10^6

Table 4.2: Performance indices for the fetal ECG extraction from the ECG recordings shown in figure 4.7. In parentheses, values in dB.

References

[1] E. Bacharakis. Separation of Maternal and Foetal ECG using Blind Source Separation Methods. Master's thesis, University of Strathclyde, Glasgow, Scotland, UK, September 1995.

[2] E. Bacharakis, A. K. Nandi, and V. Zarzoso. Foetal ECG Extraction using Blind Source Separation Methods. In *Proceedings EUSIPCO'96*, pages 395–398, Trieste, Italy, 10th–13th September 1996.

[3] A. J. Bell and T. J. Sejnowski. An Information-Maximization Approach to Blind Separation and Blind Deconvolution. *Neural Computation*, 7(6):1129–1159, 1995.

[4] A. Belouchrani, K. Abed-Meraim, J.-F. Cardoso, and E. Moulines. A Blind Source Separation Technique using Second-Order Statistics. *IEEE Transactions on Signal Processing*, SP-45(2):434–444, February 1997.

[5] A. Belouchrani and J.-F. Cardoso. Maximum Likelihood Source Separation for Discrete Sources. In *Proceedings EUSIPCO'94*, pages 768–771, Edinburgh, Scotland, UK, September 1994.

[6] E. Biglieri and K. Yao. Some Properties of Singular Value Decomposition and their Applications to Digital Signal Processing. *Signal Processing*, 18(3):277–289, November 1989.

[7] D. R. Brillinger. *Time Series. Data Analysis and Theory*. Holden-Day Inc., San Francisco, 1981.

[8] D. Callaerts, B. D. Moor, J. Vandewalle, W. Sansen, G. Vantrappen, and J. Janssens. Comparison of SVD Methods to Extract the Foetal Electrocardiogram from Cutaneous Electrode Signals. *Medical & Biological Engineering & Computing*, 28:217–224, May 1990.

[9] X.-R. Cao and R. Liu. General Approach to Blind Source Separation. *IEEE Transactions on Signal Processing*, SP-44(3):562–571, March 1996.

[10] J.-F. Cardoso. Blind Identification of Independent Components with Higher-Order Statistics. In *Proceedings Workshop on Higher-Order Spectral Analysis*, pages 157–160, Vail, Colorado, June 1989.

[11] J.-F. Cardoso. Source Separation using Higher-Order Moments. In *Proceedings ICASSP'89*, pages 2109–2112, Glasgow, Scotland, UK, 23rd–26th May 1989.

[12] J.-F. Cardoso. On the Performance of Orthogonal Source Separation Algorithms. In *Proceedings EUSIPCO'94*, pages 776–779, Edinburgh, Scotland, UK, September 1994.

[13] J.-F. Cardoso. The Invariant Approach to Source Separation. In *Proceedings NOLTA'95*, pages 55–60, 1995.

[14] J.-F. Cardoso. Performance and Implementation of Invariant Source Separation Algorithms. In *Proceedings ISCAS'96*, 1996.

[15] J.-F. Cardoso. Infomax and Maximum Likelihood in Blind Source Separation. *IEEE Signal Processing Letters*, 4(4):112–114, April 1997.

[16] J.-F. Cardoso. Statistical Principles of Source Separation. In *Proceedings IFAC SYSID'97*, pages 1837–1844, Fukuoka, Japan, 1997.

[17] J.-F. Cardoso and B. H. Laheld. Equivariant Adaptive Source Separation. *IEEE Transactions on Signal Processing*, SP-44(12):3017–3030, December 1996.

[18] J.-F. Cardoso and A. Souloumiac. Blind Beamforming for non-Gaussian Signals. *IEE Proceedings-F*, 140(6):362–370, December 1993.

[19] E. Chaumette, P. Comon, and D. Muller. ICA-Based Technique for Radiating Sources Estimation: Application to Airport Surveillance. *IEE Proceedings-F*, 140(6):395–401, December 1993.

[20] R. M. Clemente and J. I. Acha. Blind Separation of Sources using a New Polynomial Equation. *IEE Electronics Letters*, 33(3):176–177, 30th January 1997.

[21] P. Comon. Separation of Sources using Higher-Order Cumulants. In *SPIE Vol. 1152 Advanced Algorithms and Architectures for Signal Processing IV*, pages 170–181, 1989.

[22] P. Comon. Separation of Stochastic Processes. In *Proceedings Workshop on Higher-Order Spectral Analysis*, pages 174–179, Vail, Colorado, June 1989.

[23] P. Comon. Statistical Approach to the Jutten-Herault Algorithm. In *Proceedings NATO Workshop on Neuro-Computing*, Les Arcs, France, 27th February – 3rd March 1989. Republished in: *Neurocomputing. Algorithms, Architectures and Applications*, F. Fogelman and J. Herault (Eds.), NATO ASI Series, pp. 81–88, Springer Verlag, 1990.

[24] P. Comon. Independent Component Analysis, A New Concept? *Signal Processing*, 36(3):287–314, April 1994.

[25] P. Comon. Tensor Diagonalization, A Useful Tool in Signal Processing. In *Proceedings IFAC SYSID'94*, pages 77–82, Copenhagen, July 1994.

[26] P. Comon, C. Jutten, and J. Herault. Blind Separation of Sources, Part II: Problems Statement. *Signal Processing*, 24(1):11–20, November 1991.

[27] T. M. Cover and J. A. Thomas. *Elements of Information Theory*. John Wiley & Sons, Inc., New York, 1991.

[28] M. Gaeta and J.-L. Lacoume. Sources Separation without a priori Knowledge: the Maximum Likelihood Solution. In *Proceedings EU-SIPCO'90*, pages 621–624, Barcelona, 1990.

[29] G. H. Golub and C. F. V. Loan. *Matrix Computations*. The Johns Hopkins University Press, Baltimore, Maryland, 2nd edition, 1989.

[30] F. Harroy and J.-L. Lacoume. Maximum Likelihood Estimators and Cramer-Rao Bounds in Source Separation. *Signal Processing*, 55:167–177, December 1996.

[31] F. Harroy, J.-L. Lacoume, and M. A. Lagunas. A General Adaptive Algorithm for nonGaussian Source Separation without any Constraint. In *Proceedings EUSIPCO'94*, pages 1161–1164, Edinburgh, Scotland, UK, September 1994.

[32] R. A. Johnson and D. W. Wichern. *Applied Multivariate Statistical Analysis*. Prentice Hall, Englewood Cliffs, New Jersey, 2nd edition, 1988.

[33] C. Jutten and J. Herault. Une Solution Neuromimétique au Problème de Séparation de Sources. *Traitement du Signal*, 5(6):389–404, 1989.

[34] C. Jutten and J. Herault. Blind Separation of Sources, Part I: An Adaptive Algorithm Based on Neuromimetic Architecture. *Signal Processing*, 24(1):1–10, November 1991.

[35] J.-L. Lacoume and P. Ruiz. Sources Identification: A Solution Based on the Cumulants. In *Proceedings IEEE ASSP Workshop*, Minneapolis, August 1988.

[36] J.-L. Lacoume and P. Ruiz. Separation of Independent Sources from Correlated Inputs. *IEEE Transactions on Signal Processing*, SP-40(12):3074–3078, December 1992.

[37] B. H. Laheld and J.-F. Cardoso. Adaptive Source Separation with Uniform Performance. In *Proceedings EUSIPCO'94*, pages 183–186, Edinburgh, Scotland, UK, September 1994.

[38] L. D. Lathauwer, D. Callaerts, B. D. Moor, and J. Vandewalle. Fetal Electrocardiogram Extraction by Source Subspace Separation. In *Proceedings IEEE/ATHOS Signal Processing Conference on Higher-Order Statistics*, pages 134–138, Spain, June 1995.

[39] L. D. Lathauwer, B. D. Moor, and J. Vandewalle. Blind Source Separation by Higher-Order Singular Value Decomposition. In *Proceedings EUSIPCO'94*, pages 175–178, Edinburgh, Scotland, UK, September 1994.

[40] A. Mansour and C. Jutten. Fourth-Order Criteria for Blind Sources Separation. *IEEE Transactions on Signal Processing*, SP-43(8):2022–2025, August 1995.

[41] A. Mansour and C. Jutten. A Direct Solution for Blind Separation of Sources. *IEEE Transactions on Signal Processing*, SP-44(3):746–748, March 1996.

[42] K. V. Mardia, J. T. Tent, and J. M. Bibby. *Multivariate Analysis*. Academic Press, London, 1979.

[43] P. McCullagh. *Tensor Methods in Statistics*. Monographs on Statistics and Applied Probability. Chapman and Hall, London, 1987.

[44] J. M. Mendel. Tutorial on Higher-Order Statistics (Spectra) in Signal Processing and System Theory: Theoretical Results and Some Applications. *Proceedings of the IEEE*, 79(3):278–305, March 1991.

[45] A. K. Nandi and V. Zarzoso. Fourth-Order Cumulant Based Blind Source Separation. *IEEE Signal Processing Letters*, 3(12):312–314, December 1996.

[46] A. K. Nandi and V. Zarzoso. Foetal ECG Separation. In *IEE Colloquium on the Use of Model Based Digital Signal Processing Techniques in the Analysis of Biomedical Signals*, pages 8/1–8/6, Savoy Place, London, 16th April 1997.

[47] C. L. Nikias and A. P. Petropulu. *Higher-Order Spectra Analysis. A Nonlinear Signal Processing Framework.* Signal Processing Series. Prentice Hall, Englewood Cliffs, New Jersey, 1993.

[48] T. Oostendorp. *Modelling the Fetal ECG.* PhD thesis, K. U. Nijmegen, The Netherlands, 1989.

[49] D. T. Pham. Blind Separation of Instantaneous Mixture of Sources via an Independent Component Analysis. *IEEE Transactions on Signal Processing*, SP-44(11):2768–2779, November 1996.

[50] D. T. Pham, P. Garat, and C. Jutten. Separation of a Mixture of Independent Sources through a Maximum-Likelihood Approach. In *Proceedings EUSIPCO'92*, pages 771–774, Brussels, 1992.

[51] R. Plonsey. *Bioelectric Phenomena.* McGraw-Hill, New York, 1969.

[52] R. Plonsey and R. C. Barr. *Bioelectricity: A Quantitative Approach.* Plenum Press, New York, 1988.

[53] R. Roy and T. Kailath. ESPRIT - Estimation of Signal Parameters via Rotational Invariance Techniques. *IEEE Transactions on Acoustics, Speech, and Signal Processing*, ASSP-37(7):984–995, July 1989.

[54] L. L. Scharf. *Statistical Signal Processing. Detection, Estimation and Time Series Analysis.* Addison-Wesley, Inc., 1991.

[55] R. O. Schmidt. Multiple Emitter Location and Signal Parameter Estimation. *IEEE Transactions on Antennas and Propagation*, AP-34(3):276–280, March 1986.

[56] E. Sorouchyari. Blind Separation of Sources, Part III: Stability Analysis. *Signal Processing*, 24(1):21–29, November 1991.

[57] A. Stuart and J. K. Ord. *Kendall's Advanced Theory of Statistics*, volume I. Edward Arnold, London, 6th edition, 1994.

[58] L. Tong, Y. Inouye, and R. Liu. Waveform-Preserving Blind Estimation of Multiple Independent Sources. *IEEE Transactions on Signal Processing*, SP-41(7):2461–2470, July 1993.

[59] L. Tong, R. Liu, V. C. Soon, and Y.-F. Huang. Indeterminacy and Identifiability of Blind Identification. *IEEE Transactions on Signal Processing*, SP-38(5):499–509, May 1991.

[60] L. Tong, S. Yu, Y. Inouye, and R. Liu. A Necessary and Sufficient Condition of Blind Identification. In *Proceedings International Signal Processing Workshop on Higher-Order Statistics*, pages 261–264, Chamrousse, France, 10th–12th July 1991.

[61] J. Vanderschoot, D. Callaerts, W. Sansen, J. Vandewalle, G. Vantrappen, and J. Janssens. Two Methods for Optimal MECG Elimination and FECG Detection from Skin Electrode Signals. *IEEE Transactions on Biomedical Engineering*, BME-34(3):233–243, March 1987.

[62] E. Weinstein, M. Feder, and A. V. Oppenheim. Multi-Channel Signal Separation by Decorrelation. *IEEE Transactions on Speech and Audio Processing*, SAAP-1(4):405–413, October 1993.

[63] D. Yellin and E. Weinstein. Criteria for Multichannel Signal Separation. *IEEE Transactions on Signal Processing*, SP-42(8):2158–2168, August 1994.

[64] D. Yellin and E. Weinstein. Multichannel Signal Separation: Methods and Analysis. *IEEE Transactions on Signal Processing*, SP-44(1):106–118, January 1996.

[65] V. Zarzoso and A. K. Nandi. The Potential of Decorrelation in Blind Separation of Sources Based on Cumulants. In *Proceedings ECSAP'97*, pages 293–296, Prague, Czech Republic, 24th–27th June 1997.

[66] V. Zarzoso and A. K. Nandi. Generalization of a Maximum-Likelihood Approach to Blind Source Separation. In *Proceedings EUSIPCO'98*, volume IV, pages 2069–2072, Rhodes, Greece, 8th–11th September 1998.

[67] V. Zarzoso and A. K. Nandi. Modelling Signals of Arbitrary Kurtosis for Testing BSS Methods. *IEE Electronics Letters*, 34(1):29–30, 8th January 1998. Errata: Vol. 34, No. 7, 2nd April 1998, p. 703.

[68] V. Zarzoso, A. K. Nandi, and E. Bacharakis. Maternal and Foetal ECG Separation using Blind Source Separation Methods. *IMA Journal of Mathematics Applied in Medicine & Biology*, 14:207–225, 1997.

5 ROBUST CUMULANT ESTIMATION

D Mämpel and A K Nandi

Contents

5.1	Introduction	254
5.2	AGTM, LMS and LTS	255
	5.2.1 Measure of tail lengths	256
	5.2.2 Measure of asymmetry	256
5.3	The $q_0 - q_2$ plane	257
5.4	Continuous probability density functions	257
5.5	Algorithm	259
	5.5.1 Estimation of q_0 and q_2	259
	5.5.2 Truncation parameters	260
	5.5.3 Segmentation of the q_0-q_2 plane	260
	5.5.4 Summary of the algorithm	262
5.6	Simulations and results	263
	5.6.1 AM, LMS, LTS, AGTM — system input	263
	5.6.2 AM, LMS, LTS, AGTM — system output	263
	5.6.3 Effect of the amount of data	265
	5.6.4 Third–order cumulants	267
	5.6.5 Autocorrelation	270
5.7	Concluding Remarks	273
	References	275

5.1   Introduction

One of the problems in the application of higher-order statistics (HOS) is that
of the estimation of cumulants. The higher the order the larger tends to be
the variance in the estimated cumulants and this problem is also enhanced by
the limited number of samples used in applications. Naturally, the accuracy
of the methods based on higher–order statistics depend on, among other
things, the consistency of the estimates of the cumulants. Some aspects of
HOS estimators [2, 15] are not followed up here.

The standard estimation method employs the *arithmetic mean (AM)* estimator. The third order cumulant sequence of a zero-mean stationary time-
series $x(k)$ can be defined by $C_3(m,n) = \mathrm{E}[x(k)x(k+m)x(k+n)]$, where $\mathrm{E}[\cdot]$
is the expectation operator and k is the time index. The standard estimation
method evaluates

$$\hat{C}_3(m,n) = \frac{1}{K} \sum_{k=1}^{N-m} x(k)x(k+m)x(k+n) , \qquad (5.1)$$

for $m \geq n \geq 0$, where $K = N$ represents a biased estimator and $K = (N-m)$
represents an unbiased estimator. By defining $z_k = x(k)x(k+m)x(k+n)$
for fixed values of m and n,

$$\hat{C}_3(m,n) = \frac{1}{K} \sum_{k=1}^{N-m} z_k , \qquad (5.2)$$

which is equivalent to using the AM estimator. Often the time-series data are
segmented and the required cumulant in each of these segments is estimated
separately using the mean estimator, and then for the final estimate the mean
of these estimates is calculated over all the segments [21].

Recently a number of robust estimators of location parameters have been
used, in simulations, to estimate the third order cumulants of various lags
[17]. For symmetric distributions in z_k, a number of robust L- and M-
estimators, like the median, biweight, and wave estimators (weighted or un-
weighted), can provide cumulant estimates. These robust estimators are non-
adaptive and were designed for symmetric distributions [3, 8–10, 12, 13, 22, 29]
for which the mean, median, and centre of symmetry are all the same. For
asymmetric probability density functions in z_k the various location parameters like the mean, median, etc. are different and naturally the aforementioned estimators do not work for the estimation of the mean; this now raises
the question of what should be an alternative mean estimator. Therefore,
for skewed distributions, it is particularly important to define the location
parameter to be estimated [6, 12].

Herein particular emphasis is put on the cumulant estimation from skewed distributions [16, 18, 20]. A variety of short and long tailed symmetric, right skewed, and left skewed distributions have been studied in simulations. Instead of treating symmetric distributions separately from skewed distributions, one theme of this work has been to look for a unified approach to the estimation from both symmetric and skewed distributions. Various estimators, including winsorized mean [11], trimmed mean, generalised trimmed mean [11, 23], and bootstrap approaches [22, 31], have been explored though not all are reported here. In this chapter is detailed the *adaptive generalised trimmed mean (AGTM)* estimator which allows the possibility of truncating a different number of samples from the left side tail and the right side tail. The performance of the AGTM is compared with those of AM, *least median of squares (LMS)* [23–25], and *least trimmed squares (LTS)* [23–25].

5.2   AGTM, LMS and LTS

Let us consider that there are N data samples, $\{z_i, i = 1, 2, ..., N\}$ from which we make up a set of these N samples in the ascending order $\{Z_i, i = 1, 2, ...N\}$ such that $Z_1 \leq Z_2 \leq \cdots \leq Z_N$. For asymmetric discrete data samples, the AGTM is defined [11, 23] as

$$m_T(\alpha_\mathrm{L}, \alpha_\mathrm{R}) = \frac{1}{N - g_\mathrm{L} - g_\mathrm{R}} \sum_{i=1+g_\mathrm{L}}^{N-g_\mathrm{R}} Z_i \,, \tag{5.3}$$

where $g_j = \alpha_j N$ if $\alpha_j N$ is an integer, else $g_j = [\alpha_j N]$, for $j = \mathrm{L}$ or R, with $[\cdot]$ being the greatest integer function. The AGTM excludes, in general, a different number of data samples from the left (g_L) and right (g_R) tails of a distribution. Hence, this is suitable for asymmetric distributions if α_L and α_R can be chosen such that the AGTM, $m_T(\alpha_\mathrm{L}, \alpha_\mathrm{R})$, is an unbiased estimator. This choice needs to be adaptive, in that the values of α_L and α_R must be data dependent and it must reflect the tail lengths and asymmetry. In the next section these concepts are explored.

Results from the AM estimator and two of the robust estimators — namely the LMS [23–25] and the LTS [23, 25] — are compared with those from the proposed AGTM estimator. Unlike the least squares method which chooses the least sum of squares the LMS yields the least median of squares and the use of the median makes it less susceptible to outliers. Indeed its breakdown point is 50%. The LMS and LTS estimates can be defined as

$$r_i^2(m) = (Z_i - m)^2 \,, \qquad \text{for} \qquad i = 1, 2, \ldots, N \tag{5.4}$$

and $\{R_i^2(m)\}$ is a set of N values of r_i^2 in the ascending order such that $R_1(m) \leq R_2(m) \leq \cdots \leq R_N(m)$:

$$\hat{m}_{\text{LMS}} = \underset{m}{\text{Minimise median}} R_i^2(m)$$

$$\hat{m}_{\text{LTS}} = \underset{m}{\text{Minimise}} \sum_{i=1}^{h} R_i^2(m) \qquad (5.5)$$

In LTS, for each choice of m, the sum of the smallest h squared residuals are calculated. The LTS estimate, $\hat{m}_{\text{LTS}}$, corresponds to that value of m that gives rise to the smallest sum of the first h squared residuals. The best robustness properties are achieved when h is approximately $N/2$, in which case the breakdown point reaches 50%.

5.2.1 Measure of tail lengths

Tail lengths of a distribution can be measured in a number of ways [11]. The following measure is considered in this chapter:

$$q_0 = \frac{\bar{U}(0.05) - \bar{L}(0.05)}{\bar{U}(0.5) - \bar{L}(0.5)}, \qquad (5.6)$$

where $\bar{U}(\alpha)$ and $\bar{L}(\alpha)$ are the averages of the largest $g = [\alpha N]$ order statistics and of the smallest g order statistics respectively. For discrete data samples, they can be estimated by

$$\bar{U}(\alpha) = \left(\sum_{j=N-g+1}^{N} Z_i \right) \bigg/ g \quad \text{and} \quad \bar{L}(\alpha) = \left(\sum_{i=1}^{g} Z_i \right) \bigg/ g . \qquad (5.7)$$

It should be noted that the q_0, being the ratio of two linear functions of order statistics, is a robust measure. The Gaussian distribution ($q_0 = 2.58$) has a shorter tail than the Laplace (or double exponential) distribution ($q_0 = 3.30$), while the value of q_0 corresponding to the uniform distribution is 1.9. Another indicator of length of tails has sometimes been taken as the value of the kurtosis. Since the convergence properties [11] of q_0 is better than those of the kurtosis, the parameter q_0 is preferred in this work.

5.2.2 Measure of asymmetry

Like the tail lengths of a distribution, a number of measures of asymmetry are available [11,27]. In this chapter the following measure is used:

$$q_2 = \frac{\bar{U}(0.05) - m(0.25)}{m(0.25) - \bar{L}(0.05)}, \qquad (5.8)$$

where one can define for discrete data samples, the α-trimmed mean of these N samples as

$$m(\alpha) = \frac{1}{N - 2g} \sum_{i=1+g}^{N-g} Z_i \qquad (5.9)$$

where $g = [\alpha N]$, with $[\cdot]$ being the greatest integer function. Distributions with $q_2 > 1$ have comparatively longer right tails, while those with $0 < q_2 < 1$ have comparatively longer left tails. The value of q_2 lies in the range zero to infinity and it is equal to one for symmetric distributions.

5.3 The $q_0 - q_2$ plane

Any unimodal and regular distribution can now be mapped on to the $q_0 - q_2$ plane, where q_0 is the x-axis and the q_2 is the y-axis. These parameters are ratios of differences of averages and naturally all these differences are independent of any shift of the density functions; therefore the parameters themselves are independent of any shift of the density functions. Furthermore the q_0 and q_2 are independent of any scale parameter that is used in pdfs. On this account, both parameters can be regarded as *invariants* [27]. Hence all investigations of the known continuous pdfs are carried out with zero shift and scale parameter equal to one.

The relationship between *longer right tail (lrt)* and *longer left tail (llt)* distributions can be considered by choosing any llt distribution $p(z)$. The asymmetry of such distributions is described by $q_{2llt} < 1$. If such a distribution is reflected about any value of z, a lrt distribution will be obtained and the asymmetry of the reflected distribution will be described by $q_{2lrt} = 1/q_{2llt}$. Thus both the llt and right tail distributions are symmetrically distributed on a logarithmic q_2-axis with symmetric distributions corresponding to zero value in the middle of this logarithmic axis. In the following work we focus on the lrt distributions without any loss of generality since llt distributions can be transformed to lrt distribution with the unique inversion property.

5.4 Continuous probability density functions

The current approach is to estimate the parameters — q_0 and q_2 — from the available discrete data samples whose underlying pdf is not known and then to find theoretical pdfs that are good approximations. For this reason, q_0 and q_2 of many pdfs — e.g. symmetric distributions like uniform, Gaussian, Laplace, Cauchy, Rayleigh, generalised Gaussian as well as asymmetric distributions like exponential, lognormal, Gamma, Beta, Weibull, Maxwell-Boltzman, etc. [1,5,7] — have been computed. It is found that these distributions cover a small area of the $q_0 - q_2$ plane compared to the range of the data and therefore they are unsuitable in describing most of the data.

Recently there have been interests in the α-*stable pdf* within the signal processing community (for example see [26,28] and references therein). In [28] estimation methods of parameters of symmetric α-stable (SαS) impulsive interference are described. A SαS pdf is completely characterised by three parameters: $0 <$ characteristic exponent $(\alpha) \leq 2$, dispersion > 0, and $-\infty <$ location parameter $< \infty$. The SαS pdf generates heavier tails than the Gaussian distribution and the characteristic exponent relates directly to the heaviness of these tails (smaller values of this exponent produces heavier tails). The value of $\alpha = 2$ corresponds to the Gaussian distribution while the value of $\alpha = 1$ refers to a Cauchy distributions. As all the SαS pdfs are symmetric these are of little use in describing asymmetric distributions. There are of course other α-stable distributions which are asymmetric [27]. It can be shown that any α-stable distribution is infinitely divisible [27] and it is known that the fourth-order cumulant is ≥ 0 for all infinitely divisible distributions [4]. Therefore no α-stable distribution can describe any pdf whose fourth order cumulant is < 0 (for example, the uniform distribution).

For this purpose, the *extended generalised Gaussian distribution* [19] is considered. This is an extension of the generalised Gaussian distribution [30] which can offer different length of tails but always remains symmetric. The extended generalised Gaussian distribution, which can describe symmetric as well as asymmetric pdfs, is defined [19] as

$$p(z; \gamma_n, \gamma_p) = \frac{\gamma_n}{2\Gamma(1/\gamma_n)} \exp\left\{-(-z)^{\gamma_n}\right\}, \qquad \text{for} \qquad z \leq 0, \qquad (5.10)$$

and

$$p(z; \gamma_n, \gamma_p) = \frac{\gamma_n}{2\Gamma(1/\gamma_n)} \exp\left\{-\left(\frac{\gamma_n\Gamma(1/\gamma_p)z}{\gamma_p\Gamma(1/\gamma_n)}\right)^{\gamma_p}\right\} \qquad \text{for} \qquad z \geq 0, \qquad (5.11)$$

for $\gamma_n > 0$, and $\gamma_p > 0$. This pdf satisfies the required property that

$$\int_{-\infty}^{\infty} p(z; \gamma_n, \gamma_p)dz = 1, \qquad (5.12)$$

for $\gamma_n > 0$ and $\gamma_p > 0$. For all allowable values of γ_n and γ_p, the following two observations are noted: 1) for all choices of $\gamma_n = \gamma_p$, only symmetric distributions are obtained — all those that can be obtained from the generalised

Gaussian distribution and 2) for all choices of $\gamma_n \neq \gamma_p$, one obtains only asymmetric distributions — those that cannot be obtained from the generalised Gaussian distribution. By varying the values of the two parameters γ_n and γ_p we can very effectively span the q_0–q_2 plane.

Throughout this study unimodal distributions are under consideration. The plan is to estimate the q_0 and q_2 from the data. These values of the q_0 and q_2 correspond to specific and known values of the γ_n and γ_p which define one specific extended generalised Gaussian distribution. In the next section the estimation of q_0, q_2, and the truncation parameters are described.

5.5 Algorithm

5.5.1 Estimation of q_0 and q_2

Suppose R sets of data samples are available from each of the R realisation of the same process. Since they all describe the same process there is merit in using the same truncation parameters α_L and α_R for each of these R sets. The exact values of these truncation parameters come from the knowledge of q_0 and q_2, as well as the approximating extended generalised Gaussian distribution. If one calculates the values of q_0 and q_2 in each realisation, one would obtain R different values of each. To describe the effective length of tails of R realisations, the AM of the R values of q_0 is used:

$$q_0 = \frac{1}{R} \sum_{i=1}^{R} q_0(i) \,. \tag{5.13}$$

Llt distributions have $0 < q_2 < 1$ while lrt distributions have $1 < q_2 < \infty$. All symmetric distributions correspond to $q_2 = 1$ but specific realisations from them can have q_2 values on either side of 1. Of course llt distributions are mirror symmetric to lrt distributions but the parameter q_2 does not reflect that. A better parameter to highlight this observation would be the the logarithm of q_2, say $\ln(q_2)$, which would have the range of $-\infty$ to 0 for llt distributions and the range of 0 to ∞ for lrt distributions. The arithmetic mean of $\ln(q_2)$ is equivalent to the natural logarithm of the geometric mean of q_2. Therefore, the effective value of the asymmetry describing the R different realisations is obtained from the geometric mean of the R values of q_2:

$$q_2 = \left(\prod_{i=1}^{R} q_2(i) \right)^{1/R} \,. \tag{5.14}$$

As has been explained above, llt distributions can be transformed to lrt distribution by inverting the value of q_2, and in the following only the case of lrt distributions is considered without any loss of generality.

5.5.2 Truncation parameters

Using equations (5.13) and (5.14), only one value for each of q_0 and q_2 is computed to represent R sets of data samples. However to employ the proposed AGTM, the relevant truncation parameters — α_L and α_R — must be determined. At this stage, the extended generalised Gaussian distribution is used to approximate the observed cumulant distribution. The values of α_L and α_R are to be found such that the corresponding extended generalised Gaussian distribution possesses identical values of q_0 and q_2 as estimated from the data. As there are two free parameters — α_L and α_R, any one of these can be fixed and the other can be determined. It is suggested that the truncation parameter of the shorter left tail end, α_L, is held fixed and α_R, which is generally smaller than α_L and increasing with decreasing q_0 and q_2, is computed for a given pair of values of q_0 and q_2. For chosen q_0, q_2, and fixed α_L, the value of α_R is computed to generate a table of α_L, q_0, q_2, and α_R. It is necessary to produce this table only once.

5.5.3 Segmentation of the q_0–q_2 plane

Simulation results show that for the variety of distributions of data a large area of the q_0–q_2 plane is covered. Using the same fixed truncation parameter α_R for the whole q_0–q_2 plane leads to a wide range of the α_L. To overcome the restriction to truncate a significant number of samples but not too many samples at both sides of the distribution, the q_0–q_2 plane is split up into different regions with different fixed truncation parameters α_R (figure 5.1). The areas of the q_0–q_2 plane have been chosen in the following way. If the fixed α_R in one area by extending q_0 and/or q_2 leads to an α_L that is greater than 30% then the value of α_R is reduced. The borders of the areas have been approximated by exponentials of polynomial functions. Furthermore the q_0–q_2 plane has been split up into two general parts in the q_0 direction, the shorter tail and longer tail part. The segmented q_0–q_2 plane is given in figure 5.1 containing nine regions. The capital letters identify the regions and the percentage figure is the set value of α_R in that region. If the value of q_2 is less than 1.1 (regions E and I), the distribution is deemed to be symmetric and the α_L and the α_R are set equal to each other. The values of the parameters q_0 and q_2 from the data are calculated from the data. To obtain the theoretical p.d.f. to correspond to this (q_0–q_2), a two-dimensional

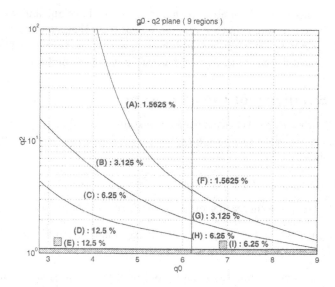

Figure 5.1: Segmentation of the q_0–q_2 plane into 9 regions.

iterative Newton algorithm is used to calculate the parameters γ_n and γ_p. Given the parameters of the asymmetric generalised Gaussian distribution and the fixed truncation parameter (α_R), the other truncation parameter (α_L) can be estimated by using another iterative Newton procedure. In this way this algorithm needs the application of two iterative procedures to obtain α_R. Therefore this is avoided to reduce the computational complexity.

The first stage is removed by the generation of table interpolation values which has to be done only once and for all, and consequently not contributing to the computational complexity of the AGTM. To ensure a better implementation of the algorithm, for each area of the q_0–q_2 plane a table of values for $\alpha_L L$ are calculated using the fixed α_R for this area. This table then is used to perform a table interpolation for the particular parameters q_0 and q_2 of the data. The accuracy of the table interpolation increases by the number of sets in the table but the computing burden increases as well. To overcome this problem and obtain a fast and accurate implementation each area of the q_0–q_2 plane can be divided into subcells, by which the same α_R is used for each subcell in the same area. Note that an overlapping of the areas and subcells is necessary to prevent interpolation errors near the borders of the areas and cells. Using table interpolation in such small cells allows a significantly faster implementation than the one of biweight and wave estimators [14], and a somewhat slower implementation than, though of similar order of, the AM estimation. Furthermore, now one can address asymmetric cumulant distributions which was not possible using estimators in [14, 17].

5.5.4 Summary of the algorithm

The algorithm for the AGTM estimator is summarised below:

Given a set of data samples in each of the R Realisations, for a cumulant of fixed lags:

1. calculate z_i's, in each realisation,

2. sort the z_i's in increasing order, in each realisation,

3. compute the values of q_0 and q_2 (using equations (5.6) and (5.8)), in each realisation,

4. calculate the arithmetic mean (q_{0AM}) of R values of q_0 as the indicator of tail lengths (using the equation (5.13)),

5. calculate the geometric mean (q_{2GM}) of R values of q_2 as the indicator of asymmetry (using the equation (5.14)),

6. if $q_{2GM} < 1$ (left tail is longer), define $q_{2GM} = 1/q_{2GM}$,

7. look up the relevant region of the q_0–q_2 plane and fix the value of α_L,

8. compute the corresponding value of α_R by table interpolation (the function 'griddata' in MATLAB has been used),

9. if the original value of q_{2GM} was less than 1 (left tail is longer), exchange α_L and α_R, and

10. compute the AGTM, $m_T(\alpha_L, \alpha_R)$.

This algorithm requires interpolations from tabulated numbers to decide the range of the data samples to accept and computing the arithmetic mean of the accepted data samples. Therefore the computational complexity of this estimator is about twice that of the AM estimator; the greater computational load related to the AGTM goes into constructing 'the table' from which the interpolation is carried out but this needs to be done only once and before any application and in sorting the samples (z_i's) in ascending order. Some simulation details and results based on the above algorithm are presented in the next section.

5.6 Simulations and results

Extensive simulations, within the MATLAB environment, have been carried out to compare the results from the AGTM estimator with those from the AM estimator, the LMS estimator, and the LTS estimator.

5.6.1 AM, LMS, LTS, AGTM — system input

Each of the following results in Table 5.1 is based on 100 simulations each of which consists of 256 i.i.d. samples generated from zero-mean symmetric and asymmetric distributions - a Gaussian distribution of variance 1, a triangular distribution in the range [-1.0,1.0], a uniform distribution in the range [-1.0,1.0], and an exponential distribution of unit variance. For each estimator, the mean and standard deviations of the estimates of three representative cumulants $C_3(1,0)$, $C_3(1,1)$, and $C_3(2,2)$ are presented in Table 5.1. Given that $\{\hat{C}_3(m,n)\}_r$ is the estimate in the r-th realisation, the mean and the standard deviation of the R estimates are defined by

$$\text{mean}\left(\hat{C}_3(m,n)\right) = \frac{1}{R}\sum_{r=1}^{R}\{\hat{C}_3(m,n)\}_r \,, \tag{5.15}$$

and

$$\text{std}\left(\hat{C}_3(m,n)\right) = \sqrt{\frac{1}{R}\sum_{r=1}^{R}\left(\{\hat{C}_3(m,n)\}_r - \text{mean}\{\hat{C}_3(m,n)\}\right)^2} \,, \tag{5.16}$$

While the LMS and LTS estimates from symmetric distributions are satisfactory, the corresponding estimates from the asymmetric (exponential) distribution are not so good. Both the AM and the AGTM estimates agree with the expectation that these cumulants are zero. The significant observation is that the standard deviations of the proposed AGTM estimator are much smaller than those of the AM estimator. It is emphasised that the above observations are true for both symmetric and asymmetric distributions in z_i's.

5.6.2 AM, LMS, LTS, AGTM — system output

Here the output cumulants from three different systems being excited by the same input have been estimated using the proposed AGTM, the AM, the LMS and the LTS estimator. Each of these estimates are based on 100 Monte Carlo runs each of which consists of 256 independent samples from

Table 5.1: Estimated cumulants (mean ± standard deviation) of various distributions. Since the samples are i.i.d. the true cumulant values at lags other than zero are all null.

Distributions	$C_3(1,0)$	$C_3(1,1)$	$C_3(2,2)$
Gaussian			
AM	-.005±.088	.006±.093	.004±.084
LMS	.006±.058	.004±.058	-.001±.063
LTS	.003±.012	0±.013	0±.012
AGTM	-.001±.027	.006±.029	0±.026
Triangular			
AM	-.001±.005	-.001±.005	0±.005
LMS	0±.004	0±.005	-.001±.004
LTS	0±.001	0±.001	0±.001
AGTM	-.001±.002	0±.002	0±.002
Uniform			
AM	0±.011	-.002±.010	0±.01
LMS	-.003±.026	.003±.024	.003±.021
LTS	-.001±.001	0±.006	0±.006
AGTM	0±.007	-.001±.007	0±.007
Exponential			
AM	.03±.175	.032±.159	-.034±.134
LMS	-.075±.047	-.088±.043	-.079±.041
LTS	-.033±.016	-.033±.016	-.034±.017
AGTM	-.001±.020	.008±.022	-.048±.022

Table 5.2: Estimated cumulants (mean $\pm$ standard deviation) of various outputs with the same set of input samples from the Gaussian distribution of unit variance and zero mean. The theoretical expected values are zero.

Systems	$C_3(0,0)$	$C_3(1,1)$	$C_3(2,2)$
AR(2)			
AM	-.031±.255	.011±.144	.008±.151
LMS	.013±.187	.015±.114	.031±.123
LTS	.005±.032	.004±.027	.006±.024
AGTM	-.009±.041	.017±.051	.002±.045
MA(2)			
AM	.206±1.59	-.161±1.365	-.09±.948
LMS	-.16±1.181	-.084±.809	-.009±.835
LTS	-.017±.207	-.026±.184	-.016±.207
AGTM	.12±.285	-.12±.389	-.067±.292
ARMA(2,2)			
AM	.517±3.088	-.19±2.402	-.221±2.002
LMS	-.088±2.19	-.109±1.147	-.01±1.367
LTS	-.044±.322	-.017±.242	-.027±.225
AGTM	.437±.464	-.143±.575	-.192±.502

a zero-mean unit variance Gaussian distribution (Table 5.2) or exponential distribution (Table 5.3). These input have been used to excite an AR system defined by $y(n) = x(n) - .64y(n-2)$ with $x(n)$ and $y(n)$ being the input and output respectively, a MA system defined by $y(n) = x(n) - 2.05x(n-1) + x(n-2)$, and an ARMA system defined by $y(n) = x(n) - 2.05x(n-1) + x(n-2) - .64y(n-2)$.

The mean and standard deviations of some representative output cumulant estimates are presented in Tables 5.2 and 5.3. The LMS and LTS estimates from the symmetric (Gaussian) distribution (in table 5.2) are satisfactory but the corresponding estimates from the asymmetric (exponential) distribution (table 5.3) are very biased and not acceptable. It should be noted that both the AM and the AGTM estimates agree and that the AGTM estimates have much smaller variance than the AM estimates.

5.6.3 Effect of the amount of data

To investigate the effects on the specific choices of the number of samples (N) in each Monte Carlo run and the number of Monte Carlo runs (nsim),

Table 5.3: Estimated cumulants (mean ± standard deviation) of various outputs with the same set of input samples from the exponential distribution of unit variance and zero mean.

Systems	$C_3(0,0)$	$C_3(1,1)$	$C_3(2,2)$
AR(2)			
Theoretical value	1.585	0	.649
AM	1.506±.743	.045±.215	.588±.35
LMS	-.076±.098	-.04±.071	-.008±.061
LTS	-.025±.017	-.018±.015	-.026±.012
AGTM	1.511±.465	.033±.063	.577±.175
MA(2)			
Theoretical value	-13.23	4.305	2
AM	-11.909±5.545	3.842±2.403	1.866±1.353
LMS	.471±.523	.202±.434	.209±.465
LTS	.139±.105	.08±.075	.073±.089
AGTM	-10.921±3.414	3.655±1.205	1.959±.721
ARMA(2,2)			
Theoretical value	-11.578	8.966	-5.302
AM	-10.349±5.225	7.852±5.1	-4.736±3.863
LMS	.545±1.21	.039±.841	.419±.748
LTS	.149±.203	.011±.157	.173±.175
AGTM	-10.865±3.435	7.374±2.53	-4.698±1.679

the MA(2) model has been excited a variable number of times by a variable number of i.i.d. samples generated from the exponential distribution of unit variance and zero mean. In each scenario, the third order cumulants of the output have been calculated. The discussion below is focused on the $C_3(2,1)$ but the trends are true for other cumulants. In Figure 5.2 are displayed respectively the averages and the standard deviations of the AM estimators (marked o) and the AGTM estimators (marked +) as a function of the number of data samples, N (from 256 to 4096), for the fixed number of Monte Carlo runs (nsim=100). The bias, the absolute difference between the estimated cumulant and the theoretical value (represented by the solid line in the Figure 5.2a) calculated from the ideal statistics of the input and the system parameters, tends to reduce as the number of samples increases. In Figure 5.2b the standard deviation is observed to decrease with increasing numbers of samples for both estimators, consistent with $N^{-1/2}$ law, and that the AGTM estimator consistently outperform the AM estimator. The averages and the standard deviations of the AM estimator (marked o) and the AGTM estimator (marked +) as a function of the number of Monte Carlo runs (from nsim=25 to 100) for the fixed number of data samples (N=1024) are shown in Figure 5.3. The theoretical value calculated from the ideal statistics of the input and the system parameters is represented by the solid line in the Figure 5.3a. These figures clearly indicate that the bias reduces as the number of Monte Carlo runs increases but the standard deviation does not appear to change much once the number of Monte Carlo runs reaches 50 particularly for the AGTM estimator, and that the AGTM estimator consistently outperform the AM estimator.

5.6.4 Third–order cumulants

Simulations have been carried out using the proposed algorithm. Estimates of third–order cumulants (TOCs) of exponential and Gaussian distributed signals as well as the outputs of the ARMA(2,2) system

$$y(n) = x(n) - 2.05x(n-1) + x(n-2) - 0.64y(n-2) \qquad (5.17)$$

driven by these signals have been performed. The system is being driven by zero-mean unit standard deviation white noise of exponential distribution (Tables 5.4 and 5.5) or of Gaussian distribution (Tables 5.6 and 5.7). Results are presented in the following tables. Tables 5.4 to 5.7 contain the mean and standard deviation of estimates obtained by employing the AM estimator and the proposed AGTM estimator. Results are given for output signals without any noise as well as contaminated with additive white Gaussian noise, corresponding to the SNR of 10 dB and 20 dB. In all Tables 50

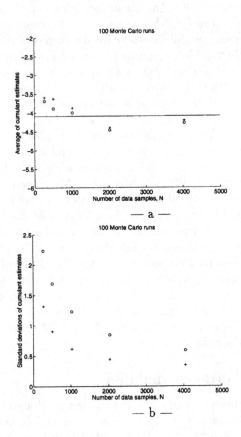

Figure 5.2: The averages (Figure 5.2a) and the standard deviations (Figure 5.2b) of the AM estimates (marked o) and the AGTM estimates (marked +) of $C_3(2,1)$ for 100 Monte Carlo runs. The input samples are taken from the exponential distribution of unit variance and zero mean. The solid line in the Figure 2a represents the theoretical value of $C_3(2,1)$.

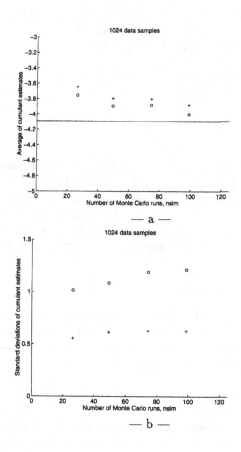

Figure 5.3: The averages (Figure 5.3a) and the standard deviations (Figure 5.3b) of the AM estimates (marked o) and the AGTM estimates (marked +) of $C_3(2, 1)$ for 1024 data samples generated from the exponential distribution of unit variance and zero mean. The solid line in the Figure 3a represents the theoretical value of $C_3(2, 1)$.

Table 5.4: TOCs (mean ± one standard deviation) of a zero-mean exponential distribution.

	AM estimates	AGTM estimates	theoretical values
$c3(0,0)$	1.851±.544	1.806±0.292	2
$c3(1,0)$	.029±.134	.014±0.015	0
$c3(2,0)$	-.04±.069	-.05±.016	0
$c3(1,1)$	.023±.113	.006±.016	0
$c3(2,1)$	.007±.036	.002±.015	0
$c3(2,2)$	-.034±.074	-.047±.017	0

realisations of the random process have been simulated using 512 samples in each realisation. For each cumulant in Tables 5.5 and 5.7, the top line of numbers corresponds to the AM estimates while the bottom line of numbers refers to the AGTM estimates. Also included in the tables are the theoretical values of the cumulants. The results show that the AGTM estimator provide cumulant estimates of less variance than the AM estimator.

5.6.5   Autocorrelation

In this application the distribution of the second–order terms becomes very asymmetric. Simulations have been carried out using the proposed algorithm to estimate the second–order cumulants of zero-mean exponential distributed signals as well as the outputs of the MA(2) system

$$y(n) = x(n) - 2.05x(n-1) + x(n-2) \qquad (5.18)$$

driven by a white zero-mean exponential input. The additive noise referred to in Table 5.9 is white and zero-mean Gaussian. Again the mean and standard deviation of the AM estimator and the AGTM estimator are given in Tables 5.8 and 5.9. For each cumulant in Table 5.9, the top line of numbers corresponds to the AM estimates while the bottom line of numbers refers to the AGTM estimates. Again included in the tables are the theoretical values of the cumulants. In Table 5.10 are presented estimates of autocorrelations from the output of the MA(2) system in eqn.(5.18) driven by a white zero-mean exponential input. The results are based on 50 Monte Carlo runs each containing 2048 samples. Four different estimators are used in the following - namely the AM, the AGTM, 1%-trimmed mean (this trims 1% from each end) and 10%-trimmed mean. It is clear that the symmetric trimmed

Table 5.5: TOCs of output of the ARMA(2,2) system driven by zero-mean exponential input.

AM estimates AGTM estimates	infinite SNR	20 dB SNR	10 dB SNR
c3(0,0)	-10.95±3.59	-10.94±3.47	-10.98±3.6
-11.58	-11.4±2.32	-11.44±2.36	-11.49±2.38
c3(1,0)	-.92±2.67	-.92±2.71	-.95±2.92
-1.43	-.69±.57	-.63±.6	-.79±.74
c3(2,0)	8.64±3.15	8.64±3.14	8.68±3.23
9.41	8.55±2	8.57±2.04	8.57±2.03
c3(1,1)	7.99±3.54	7.99±3.61	7.99±3.86
8.97	7.74±1.80	7.74±1.86	7.81±2.02
c3(2,1)	-2.94±1.54	-2.94±1.52	-2.95±1.65
-3.18	-2.85±.88	-2.83±.86	-2.86±.88
c3(2,2)	-4.7±2.63	-4.72±2.63	-4.75±2.77
-5.3	-4.5±1.12	-4.51±1.17	-4.57±1.28

Table 5.6: TOCs (mean ± one standard deviation) of a zero-mean Gaussian distribution.

	AM estimates	AGTM estimates	theoretical values
c3(0,0)	.006±.114	0±.017	0
c3(1,0)	-.002±.053	-.002±.022	0
c3(2,0)	-.017±.055	-.002±.019	0
c3(1,1)	.001±.068	-.001±.022	0
c3(2,1)	-.013±.040	-.007±.021	0
c3(2,2)	.006±.058	.002±.022	0

Table 5.7: TOCs of output of the ARMA(2,2) system driven by zero-mean Gaussian input.

AM estimates AGTM estimates	infinite SNR	20 dB SNR	10 dB SNR
c3(0,0)	.73±2.15	.79±2.14	.94±2.35
0	-.09±.27	-.10±.31	-.15±.38
c3(1,0)	-.08±1.92	-.08±1.97	-.11±2.25
0	-.07±.55	-.08±.54	-.13±..57
c3(2,0)	-.23±1.29	-.26±1.31	-.37±1.51
0	-.03±.46	-.04±.47	-.08±.54
c3(1,1)	-.44±1.87	-.53±1.86	-.73±2.06
0	-.07±.49	-.10±.49	-.18±.50
c3(2,1)	.33±1.3	.38±1.38	.49±1.67
0	.07±.55	.09±.53	.10±.55
c3(2,2)	-.30±1.34	-.3±1.38	-.27±1.63
0	-.02±.40	-.04±.37	-.08±.43

Table 5.8: Autocorrelations (mean ± one standard deviation) of a zero-mean exponential distribution.

	AM estimates	AGTM estimates	theoretical values
c2(0)	.975±.113	.977±.091	1
c2(1)	.014±.043	.040±.021	0
c2(2)	-.014±.03	.012±.022	0
c2(3)	-.007±.043	.012±.025	0

Table 5.9: Autocorrelations of the output of the MA(2) system driven by a white zero-mean exponential input.

AM estimates AGTM estimates	infinite SNR	20 dB SNR	10 dB SNR
c2(0)	5.91±.82	5.96±.83	6.47±.91
6.2	5.77±.71	5.82±.72	6.1±.77
c2(1)	-3.85±.59	-3.84±.59	-3.83±.58
-4.1	-3.76±.5	-3.75±.5	-3.75±.51
c2(2)	.88±.29	.87±.29	.86±.32
1	.82±.18	.81±.18	.81±.20
c2(3)	-.02±.32	-.02±.33	-.01±.36
0	-.01±.16	-.01±.16	-.01±.18
c2(4)	.07±.39	.07±.39	.06±.41
0	.03±.19	.03±.18	.04±.19
c2(5)	-.04±.45	-.04±.44	-.04±.44
0	.01±.2	.01±.19	0±.19

means do not offer consistent estimates while the AM and AGTM do provide consistent results.

5.7 Concluding Remarks

An adaptive generalised trimmed mean estimator for unimodal distributions has been reviewed in this chapter. It trims, in contrast to the standard symmetric trimmed mean estimator, unequal numbers of samples from the

Table 5.10: Autocorrelations of the output of the MA(2) system being excited by signals from a zero-mean exponential distribution based on 50 Monte Carlo runs with 2048 samples in each run.

	AM	AGTM	m(0.01)	m(0.10)	expected values
c2(0)	6.31±.38	6.15±.35	5.43±.31	4.08±.24	6.2
c2(1)	-4.16±.26	-4.07±.25	-3.62±.23	-2.61±.17	-4.1
c2(2)	1±.18	.97±.09	.88±.16	.65±.11	1
c2(3)	-.02±.18	.01±.08	-.01±.16	.01±.12	0

two ends of the distribution. Introducing and extracting the properties of the asymmetric generalised Gaussian distribution the number of samples to trim on the left and right side of the distribution were obtained adaptively.

Since the implementation of the AGTM makes use of table interpolations it is much faster than iterative procedures of other mean estimators, e.g. biweight and wave estimator. The AGTM has been applied to estimate second-order and third–order cumulants of various lags. A lot of simulations has been performed with symmetrically and asymmetrically distributed input and output signals from AR, MA and ARMA systems. The results show that the AGTM provides cumulant estimates consistent with expectations and of less variance than the mean estimator.

This problem of robust estimation of cumulants deserves more attention. Despite some encouraging results, this chapter contains ideas and experiments, and lacks solid theoretical foundation. More attention should be directed towards this end. This is particularly relevant either in non–stationary environments or when a small number of data samples is available for processing.

References

[1] M. Abramowitz and I. A. Stegun. *Handbook of Mathematical Functions.* Dover Publications, New York, 1972.

[2] P.-O. Amblard and J.-M. Brossier. Adaptive estimation of fourth–order cumulant of a white stochastic process. *Signal Processing*, 42:37–43, 1995.

[3] D. F. Andrews, P. J. Bickel, F. R. Hampel, P. J. Huber, W. H. Rogers, and J. W. Tukey. *Robust estimates of location: survey and advances.* Princeton Univ. Press, Princeton, NJ, 1972.

[4] R. Borges. A characterisation of the normal distribution. *Zeit. Wahrsch. Vere. Geb.*, 5:244, 1966.

[5] I. N. Bronstein and K. A. Semendjajew. *Taschenbuch der Mathematik.* BSB Teubner, Leipzig, 1989.

[6] J. R. Collins. Robust estimation of a location parameter in the presence of asymmetry. *Ann. Statist.*, 4:68–85, 1976.

[7] W. Göhler. *Höhere Mathematik.* Dt. Verl. für Grundstoffindustrie, Leipzig, 1987.

[8] F. R. Hampel. *The influence curve and its role in robust estimation*, volume 69, pages 383–393. J. Am. Stat. Assoc., 1974.

[9] F. R. Hampel, E. M. Ronchetti, P. J. Rousseeuw, and W. J. Stahel. *Robust Statistics — the approach based on influence functions.* John Wiley & Sons, New York, 1986.

[10] D. A. Hoaglin, F. Mosteller, and J. W. Tukey. *Understanding Robust and Exploratory Data Analysis.* John Wiley & Sons, New York, 1983.

[11] R. V. Hogg. Adaptive robust procedures: A partial review and some suggestions for future applications and theory. *J. Am. Statist. Assoc.*, 69:909–923, 1974.

[12] P. J. Huber. Robust statistics: A review. *Ann. Math. Statis.*, 43:1041–1067, 1972.

[13] L. A. Jaeckel. Robust estimates of location: Symmetry and asymmetric contamination. *Ann. Math. Statis.*, 42:1020–1034, 1971.

[14] G. C. W. Leung and D. Hatzinakos. Efficient implementations of higher–order statistics estimators. *Proc IEEE Signal Processing Workshop on Higher Order Statistics*, pages 266–270, Jun 1995.

[15] G. C. W. Leung and D. Hatzinakos. Implementation aspects of higher–order statistics estimators. *Journal of The Franklin Institute*, 333B(3):349–367, 1996.

[16] D. Mämpel, A. K. Nandi, and K. Schellhorn. Unified approach to trimmed mean estimation and its application to bispectrum estimation of EEG signals. *Journal of The Franklin Institute*, 333B(3):369–383, 1996.

[17] A. K. Nandi. Robust estimation of third-order cumulants in applications of higher-order statistics. *IEE Proceedings Part F*, 140(6):380–389, 1993.

[18] A. K. Nandi and D. Mämpel. On robust estimation of cumulants. In M. J. J. Holt, C. F. N. Cowan, P. M. Grant, and W. A. Sandham, editors, *Proceedings of EUSIPCO'94*, pages 470–473, 1994.

[19] A. K. Nandi and D. Mämpel. An extension of the generalised gaussian distribution to include asymmetry. *Journal of the Franklin Institute*, 332B(1):67–75, 1995.

[20] A. K. Nandi and D. Mämpel. Development of an adaptive generalised trimmed mean estimator to compute third–order cumulants. *Signal Processing*, 57:271–282, 1997.

[21] C. L. Nikias and J. M. Mendel. Signal processing with higher order spectra. *IEEE Signal Processing Magazine*, 10(3):10–37, Jul 1993.

[22] M. H. Quenouille. Notes on bias in estimation. *Biometrika*, 43:353–360, 1956.

[23] R. D. Reiss. *Approximate distributions of order statistics: with applications to nonparametric statistics*. Springer-Verlag, New York, 1989.

[24] P. J. Rousseeuw. Least median of squares regression. *J. Am. Stat. Assoc.*, 79:871–880, 1984.

[25] P. J. Rousseeuw and A. M. LeRoy. *Robust regression and outlier detection*. John Wiley & Sons, New York, 1987.

[26] M. Shao and C. L. Nikias. Signal processing with fractional lower order moments: stable processes and their applications. *Proceedings IEEE*, 81:986–1010, 1993.

[27] A. Stuart and J. K. Ord. *Kendall's Advanced Theory of Statistics*, volume I. Edward Arnold, London, 6th edition, 1994.

[28] G. A. Tsihrintzis and C. L. Nikias. Fast estimation of the parameters of alpha-stable impulsive interference. *IEEE Transactions Signal Processing*, SP-44:1492–1503, 1996.

[29] J. W. Tukey. *Exploratory Data Analysis*. Addison-Wesley, Mass., 1977.

[30] A. T. Walden. Non-gaussian reflectivity, entropy, and deconvolution. *Geophysics*, 50:2862–2888, 1985.

[31] Y. Zhang, D. Hatzinakos, and A. N. Venetsanopoulos. Bootstrapping techniques in the estimation of higher-order cumulants from short data records. In *Proceedings of ICASSP'93*, volume 4, pages 200–203, Minneapolis, 1993.

Epilogue

In the signal processing research community, a great deal of developments in higher-order statistics began in the mid-1980's. These last fifteen years have witnessed a large number of theoretical developments as well as real applications. The goal in producing this book has been to focus on the blind estimation area and to record some of these developments in this field.

Such developments are still continuing and therefore a book such as this one cannot be definitive or complete. It is hoped that the subject area has been introduced, some major developments have been recorded, and enough success as well as challenges are noted here for more people to look into higher-order statistics, along with any other information, for either generating solutions of these problems or solutions of their own problems.

Index

α-stable distribution, 258
α-trimmed mean, 257
$c(k, k)$ method, 113
$c(q, k)$ method, 111
$c(q, k + i)$ method, 113
$q_0 - q_2$ plane, 257–262
   segmentation, 260–262

adaptive/on-line processing, 208, 228–232, 242, 243
array signal processing, 169, 174
asymmetry measure, 256
autocorrelation, 6, 13
AVC93, 118

batch/block/off-line processing, 192, 193, 203, 228, 243
Bayes estimator for blind equalisation, 35
bispectrum, 10
blind channel identification, 174
blind equalisation, 42
blind signal estimation, 174
blind source separation (BSS)
   applications, 169, 170
   assumptions, 174
   convolutive mixture model, 172
   geometrical interpretation, 179, 181
   goal, 169, 174
   identifiability, 178, 241
   instantaneous linear mixture model, 173

   noiseless model, 178, **179**, 181, 192, 193, **194**, 196, 197, 207, 217, 226, 234, 240
   separability, 242
   separation principles, 242
blindness, 174
Bussgang
   Benveniste-Goursat (BG) algorithm, 47
   Bussgang equilibrium, 93
   constant modulus algorithm (CMA), 45
   Godard algorithm, 45
   Sato algorithm, 44
   stop&go (SG) algorithm, 48

central moments, 5
channel diversity, 73
characteristic function, 5
cocktail party problem, 169
code division multiple access (CDMA), 169
constant modulus algorithm, CMA, 64
contrast function, **203, 205**, 206–208, 226, 230, 241–243
   orthogonal, 230
convergence, 93
   algorithm depending equilibria, ADE, 94
   algorithms overview, 96
   Bussgang, 93
   channel depending equilibria, CDE, 94

LMS-type algorithms, 93
convolutional (residual) noise, 34
cost function, 40
criterion with memory nonlinearity, CRIMNO, 46
cumulant estimation, 128
cumulant estimators
    adaptive generalised trimmed mean (AGTM), **255**, 260–274
    algorithm, 262
    arithmetic mean (AM), **254**, 255, 259, 263–273
    biweight, 254, 262, 274
    least median of squares (LMS), **255, 256**, 263–266
    least trimmed squares (LTS), **255, 256**, 263–266
    wave, 254, 262, 274
cumulants, 4, 245
    array/tensor, **176**, 180, 182–184, 187, 208, 209, 213, 240, 245
    diagonalization, 180, **182–184**, 240–242
    cross, 175, 176, 178, 184, 187, 203, 206, 223, 240
    estimation, 7, 16
    marginal, 176, 184, 187, 205–207, 220, 223
    properties, 6
    theoretical values, 15
cyclic autocorrelation, 20
cyclo-stationary signals, 19
cyclo-statistics, 19
    equalisation alg. overview, 88
    equalisation algorithm, 84

DCQK, 148
decision directed (DD) algorithm, 47
double $c(q, k)$ method, 141

ECG recording, 189

eigen-matrices, 240
eigenvalue decomposition (EVD), 182, **192**, 194, 195, 208, 240
eigenvalues, **192**, 194
eigenvector algorithm (EVA), 62
eigenvectors, **192**, 194
equivariance, **226**, 228, 230
ergodicity, 3
estimating equations, 242
expectation-maximization (EM), 242
extended generalised Gaussian distribution, **258**, 259, 260

factor analysis (FA), 195
fetal ECG extraction, **169**, **188**, 199, 209, 216, 226, 246
full rank condition, 78
FV93, 119

Gaussian process, 2
Givens rotation, 207, 217
global matrix, **178**, 187, 226, 228, 230, 231, 244
GM89, 116, 147
Godard algorithm, 64
gradient descent algorithm, 40

higher-order arrays, 182, **211, 213**, 240
higher-order singular value decomposition (HOSVD), 213
higher-order statistics (HOS), 171, 197, 203, 234, 235, 240, 243, 254
histogram, 13
HOS
    maximum kurtosis criterion, 59

ill-convergence, 44
independence
    mutual, **175, 176**, 178, 187, 203, 205, 206, 214, 215

pairwise, 206, 207
independent component analysis (ICA), 171, **179**, **203**, 241
  direct algebraic solutions, 240, 241
  equivariant adaptive separation via independence (EASI) method, 228, 230–232, 242
  extended fourth-order blind identification (EFOBI) method, 241
  extended maximum-likelihood (EML) method, 221–227, 235, 236, 245, 246
  extended maximum-likelihood (EML) method, 217
  fourth-order blind identification (FOBI) method, 240, 241
  higher-order eigenvalue decomposition (HOEVD) method, 203, 208, 209, 212, 216, 217, 226, 230, 235, 236, 240, 245, 246
  higher-order singular value decomposition (HOSVD) method, 211, 213, 214, 216–218, 235, 236, 240, 245, 246
  infomax, 243
  joint approximate diagonalization of eigen-matrices (JADE) method, 240
  neural networks, 236
  polyspectra, 243
information theory, 205, 243
inter symbole interference, ISI, 33

Jacobi diagonalization, 208

Kullback-Leibler divergence, **205**, 241

kurtosis, 207, 221–224, 230, 241, 243, 256
  leptokurtic, long-tailed, super-Gaussian distributions, 223
  modelling, 243
  platykurtic/short-tailed/sub-Gaussian distributions, 223

least mean square (LMS) algorithm, 41
least squares (LS), 191
lenght and zero condition, 95
longer left tail (llt) distributions, **257**, 259, 260
longer right tail (lrt) distributions, **257**, 259, 260

maximum likelihood (ML), 191, 208, 222, 224, 242, 243
maximum phase, MXP, 14
mean, 3
mean square error, MSE, 33
minimum phase, MP, 14
mixing/transfer matrix, **173**, 174, 180, 195, 241, 244
MN96, 119, 129
modulation
  CCITT V.29, 39
  quadrature amplitude modulation, QAM, 38
moments, 4
  1st-order, 3
  2nd-order, 6
  3rd-order, 6
  4th-order, 6
  cyclic, 22
multi-input multi-output channel, 172
multiple channels and fractional-time spacing, 67
multiple sensor arrays, 69

mutual information, 205

NKSK95, 117
noise, 171, 175, 244
non-minimum phase, NMP, 14

observations, **169**, **171**, 244
order selection, 127

pdf expansions
   Edgeworth, 205, 222
   Gram-Charlier, 222, 242
performance indices, 185
   crosstalk (ε^2), **185**, 234, 235,
      237, 238, 245
   degree of independence (Υ), **187**,
      209, 234–237, 239, 245, 246
   flops, **187**, 234, 235, 238, 239,
      245, 246
   mixing matrix gap (ϵ), **186**, 234,
      237, 245
   rejection rate (ϱ), interference-
      to-signal ratio (ISR), inter-
      symbol interference (ISI), **186**,
      234, 235, 237, 238, 245
periodogram, 3
polar decomposition, 196
power spectrum, 3, 10
principal component analysis (PCA),
      171, **179**, 191, 194, 195, 199,
      202, 204, 207–209, 211, 214–
      216, 234–236, 245, 246
   algorithm for multiple unknown
      signals extraction (AMUSE),
      241, 242
   indeterminacy, 195–199, 203
   second-order blind identification
      (SOBI) algorithm, 242
   via EVD, 192
   via SVD, 192
probability density function, 2

Q-slice, 147
Q-slice method, 140
quasiidentity matrix, **177**, 226

relative gradient, 228, **229**, 230
residual time series, RTS, 138

sampling error, 209
scatter diagram, **197**, 198, 217, **217**,
      219, 223–225
second-order statistics (SOS), 171,
      189, 191–197
separating matrix, **178**, **180**, 182,
      187, 228, 230, 241, 244
serial updating, **228**, 229
signal subspace, **179**, 196
   ECG, **188**, 211
signal-to-noise ratio (SNR), **233**, 234,
      235, 238, 239, 245
simulation
   Bussgang, 53, 54
   channels used, 50–52
   constant modulus algprithm, CMA,
      84
   eigenvector algorithm, EVA, 67,
      68, 83
   environment, 49
   fractionally-time spaced, FTS,
      82
   super exponential algorithm, SEA,
      67, 68, 82, 83
single input multiple output, SIMO,
      76
single input single output, SISO,
      73
singular value decomposition (SVD),
      192, **193**, 194, 195, 197, 199,
      202, 204, 211, 214, 215, 240
singular values, **193**, 194–196, 199,
      214
singular vectors, **193**, 195–197, 214

snapshots, 197

source directions/signatures, **179**, 180, 181, 196–198, 219, 244

source signals, **169**, **171**, 174–177, 244

spatial filtering, 180

spatial whiteness, **175**, 176, 193, 195, 196

spectra estimation, 10

   bispectra, 11

   power spectrum, 10

spectral correlation density function, SCDF, 21

spectral diversity, 87

spectrally equivalent minimum phase, SEMP, 16

stochastic process, 2

subspace decomposition, 90

super exponential algorithm, SEA, 60

T-equation, 117

tail-length measure, 256

tensor, **180**, **182**, 208, 240

   diagonalization, 180, 240, 242

tricepstrum equalisation algorithm (TEA), 56

trispectrum, 10

truncation parameters, 259–261

uniform performance, **226**, 228–230, 241

waveform-preserving equivalence, 177

whitened observations, **196**, 215, 217, 219, 220, 222, 244

whitening, **196**, 219

whitening matrix, 196

zero forcing (ZF) algorithm, 81, 88